Warfare
in
Primitive
Societies

War/Peace Bibliography Series

Richard Dean Burns, Editor

This Series has been developed
in cooperation with the
Center for the Study of Armament and Disarmament,
California State University, Los Angeles.

Songs of Protest, War & Peace

A Bibliography & Discography
   R. Serge Denisoff

Warfare in Primitive Societies

A Bibliography
   William Tulio Divale

The Vietnam Conflict

Its Geographical Dimensions, Political Traumas, &
Military Developments
   Milton Leitenberg & Richard Dean Burns

Warfare in Primitive Societies

A Bibliography

William Tulio Divale

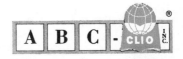

Santa Barbara, California
Oxford, England

Library of Congress Catalog Card Number 73-81978
ISBN Paperbound Edition 0-87436-122-2

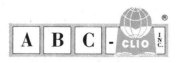

American Bibliographical Center—Clio Press, Inc.
2040 Alameda Padre Serra
Santa Barbara, California

European Bibliographical Center—Clio Press
30 Cornmarket Street
Oxford OX1 3EY, England

Designed by Barbara Monahan
Composed by Camera-Ready Composition
Printed and bound by Publishers Press
in the United States of America

To my friend and teacher KEITH F. OTTERBEIN

CONTENTS

FOREWORD

With this bibliographical series, the Center for the Study of Armament and Disarmament seeks to promote a wider understanding of martial violence and the alternatives to its employment. The Center, which was formed by concerned faculty and students in 1962-1963, has as its primary objective the stimulation of intelligent, nonpolemical discussions of war/peace issues. More precisely, the Center has undertaken two essential functions: 1) to collect and catalogue materials bearing on war/peace issues; and 2) to aid faculty, students, and members of the general public in their individual and collective probing of the historical, political, economic, philosophical, technical, and psychological facets of these fundamental problems. This bibliography series is, obviously, one tool with which we may more effectively accomplish our task.

This series is intended to provide "working," rather than definitive, bibliographies on relatively narrow themes within the spectrum of war/peace studies. This vertical approach may be supplemented by the bibliographical materials currently available. In this latter regard, we direct your attention to Blanche Cook, ed., *A Bibliography on Peace Research in History*, 1969, ABC—Clio Press, 2040 Alameda Padre Serra, Santa Barbara, California; *Arms Control and Disarmament: A Quarterly Bibliography*, 1965-, Library of Congress, Washington, D.C.; *Peace Research Review*, 1967-, and *Peace Research Abstracts*, 1964-, Canadian Peace Research Institute, Clarkson, Ontario, Canada; *Journal of Peace Research*, 1964-, International Peace Research Institute, Universitelsforlaget, University of Oslo, Box 307, Blindern, Oslo 3, Norway; *War/Peace Report*, 1961-, 8 E. 36th Street, New York, N.Y., 10016. This list is by no means complete with all the publications which deal with war/peace issues, but it will provide an ample introduction to the serious student.

While we hope this series will prove to be a useful tool, we also solicit your comments regarding improvement of its format, contents, and topics.

RICHARD DEAN BURNS
Series Editor

INTRODUCTION TO THE REVISED EDITION

In the two years since the first edition of this bibliography was published, interest and research in the problems of war and peace have increased so rapidly that a second and expanded edition was thought necessary. About four hundred new references have been added, bringing the total number of sources in this bibliography to 1655. In addition, an author index is included to supplement the society name index. I am certain that some of the initial success of this modest bibliography resulted simply because it filled a gaping void—no other bibliography of primitive warfare exists. But in these past two years there has also occurred a renewed surge in primitive war and peace research by several social scientists and their students. Fifteen years ago we knew almost nothing about the functions of primitive warfare. Today we know a great deal and our knowledge is increasing at an almost exponential rate. Even so, we have only scratched the surface and ninety percent of the needed research lies ahead. This bibliography was compiled for these scientists and their students, and I hope that it may be useful to them.

The bulk of the anthropological literature on primitive war is descriptive. A much smaller portion is theoretical; it offers hypotheses as to the causes and functions of warfare. With few exceptions the descriptive and theoretical literature follow parallel lines which only rarely interconnect. Most anthropologists describe the warfare of the people they study; and a few present general theories about warfare whose data base is generally only two or three societies. The descriptive literature grew over the years, and the theories remained untested. Now this has begun to change. Two years ago there was only a trickle of studies which presented hypotheses about the functions of war and tested those hypotheses cross-culturally or through case studies to determine if they were in harmony with the available data. That trickle has now enlarged into a steady stream and our knowledge of primitive war is continually expanding; some of these findings are discussed in this introduction. I have let the

Introduction to the First Edition stand because it still describes the main aspects of primitive war.

Keith Otterbein's recent review (no. 71) of the various theories and schools of thought about primitive warfare is the best general review of primitive war in print. Vayda (no. 99) and Levine and Campbell (no. 55) present many of the competing hypotheses about the functions of war. The research on primitive war that has been done since 1960 falls loosely into three areas: cultural evolution, postmarital residence patterns, and demography. No attempt is made in this volume to discuss all the research but I have selected what I consider to be the most important new findings. The reader should consult Otterbein's article for a more rounded view.

War & Cultural Evolution

It has long been generally believed that success in warfare and higher societal complexity (cultural evolution) were related. This view was most popular at the turn of the century and originated with the work of Herbert Spencer, William G. Sumner, and Albert Keller. Modern theorists Walter Goldschmidt, Robert Carneiro and others maintain this perspective. This theory (which represents the scientific aspect of Social Darwinism) holds that old or ineffective social practices and institutions are weeded out of society through warfare. Societies survive—i.e., those successful in warfare—because their institutions are more adaptive and they expand at the expense of weaker societies which perish. As a result of this continual process societies experience cultural evolution—e.g., they become more complex. According to the theory of Social Darwinism warfare is a mechanism of natural selection in human society. A recent cross-cultural test conducted by Naroll and Divale (no. 64) failed to support the theory.

Naroll and Divale tested the hypothesis: "if warfare is the selective mechanism of cultural evolution, then militarily successful societies should tend to be higher on the scale of cultural evolution than militarily unsuccessful societies" by examining a world wide sample of forty-nine societies ranging from bands of hunters-gatherers to modern European societies. They measured military success as a percent of territorial change, in terms of ground lost or gained by a society in one-hundred-year periods. Territorial change turned out to be a good objective measure of military success since ninety percent of all territorial change experienced by the societies in the sample was the result of warfare. Various indexes of cultural evolution were employed to measure the degree of urbanization, economic specialization, social organization, and political integration of the societies studied. The results were conclusive, *none* of the measures of cultural evolution were in any way related to military success. In another study, Otterbein (no. 69) found similar results. Thus the theory that warfare is the selective mechanism of cultural evolution must be rejected. More research is, of course, still needed on this complex problem.

War & Postmarital Residence

Recent research on the relationship between warfare and postmarital residence patterns has greatly increased our understanding of some of the functions and interactions of warfare and social structure. It is also significant that these studies were conducted by several different anthropologists working independently, yet the results are complementary and appear to fit like the parts of an engine. Residence patterns are one of the most important aspects of social structure because different patterns will result in married couples living with different groups of relatives. This has important implications for primitive societies, much more than for our own, because almost all economic, social, political, and religious activity is conducted by and through groups of either cognatic relatives (related through blood) or affinal relatives (related through marriage).

Van Velzen and van Wetering (no. 97) report that societies with matrilocal residence (wherein married couples reside with or near the wife's relatives) attempt to suppress feuds and as a whole are much more peaceful internally than are societies with patrilocal residence (where the married couple reside with or near the husband's relatives). In contrast to this, Otterbein and Otterbein (no. 350) report that societies with patrilocal residence tend to feud frequently in comparison to societies with other residence patterns where feuding tended to be absent.

In another study, Otterbein (no. 70; see also no. 69) reports that societies with patrilocal residence tend to be engaged in continual or frequent internal warfare. Internal war is defined as fighting between communities of the same society and feuding is defined as fighting between members of the same community. Otterbein also reports that polygyny (two or more women married to the same man) is associated with both internal warfare and patrilocal residence.

In contrast to this, Mel and Carol Ember (no. 28) report that societies with matrilocal residence tend to wage only external warfare [this finding was successfully replicated by Divale, (nos. 24, 25)]. External war is defined as fighting between communities of different societies, e.g., different linguistic units.

Recent research by Divale (nos. 24, 25) suggests that matrilocal residence is a response made by previously patrilocal societies to a migration into already-inhabited areas. The immigration of the patrilocal society results in overcrowding and external warfare between the migrating and the indigenous society. In this situation, a survival value would be placed on the cessation of internal warfare which characterizes those patrilocal societies. This can be accomplished by a switch to matrilocal residence which results in the breakup of localized groups of fraternal males. This is because upon marriage these males are scattered into several communities where they go to live with their wives. It is these localized groups of related males which are conducive to internal warfare.

Now, in pulling the above findings together into a single picture, it is helpful to note that about seventy percent of the approximately three thousand societies in the world practice patrilocal residence while about ten percent are matrilocal. These patrilocal societies are characterized by frequent feuding, continual or frequent internal warfare, and extensive polygyny. In other words, patrilocal societies are much more warlike than matrilocal societies. Matrilocal residence which is a result of previous migration leads to the development of internal harmony: feuds and internal warfare are suppressed and usually only external warfare is present.

War & Demography

Recent research on demography and primitive warfare has taken two directions. One trend of research, conducted primarily in New Guinea, investigates the hypothesis that warfare functions to redistribute population over available land and to maintain equitable man-land and man-resource ratios. A second trend has been to test the hypothesis that primitive warfare functions as a population control device.

Andrew Vayda (no. 98), Roy Rappaport (no. 1607), and Mervyn Meggitt (no. 62) have conducted field studies in the New Guinea Highlands on the systemic relationships between warfare, demography, and social organization. Their findings, as summarized by Vayda (no. 98), indicate that warfare in New Guinea operates in twenty-year cycles. In the New Guinea Highlands, communities are bonded into strong warfare alliances with several other communities. These alliances usually last about twenty years. Warfare consists of raiding and open pitched battles. A raid will usually result in only one death and is followed in a few weeks by a revenge raid conducted by the injured village. The open battles are almost of a ritual nature and rarely result in a death. General population growth is controlled somewhat by the constant raiding. However, in the course of time (usually twenty years or so) local demographic changes occur. Within a particular alliance, some communities may have an increase in population while others may have had a decrease. Thus there is great variation in local population pressure on the land. Communities that have had a growth in population need more land and to get it they sometimes conspire with their enemies to attack members of their own alliance. When this happens the allies are caught by surprise—slaughter and a change in territorial boundaries are the result. In this way populations are redistributed over the land.

The findings of the Vayda group are significant because the pattern of warfare which includes both raiding and open battles is found in several regions throughout the world. If this pattern is indicative of the twenty-year land redistribution cycles then it means that those findings are not limited to the New Guinea Highlands but may apply elsewhere in the world as well. Further research, especially cross-cultural testing, should be conducted in this area.

The population control functions of primitive warfare have been another area of fruitful research. It is not yet known whether modern warfare has population control functions. This is because the number of individuals killed, while large in absolute numbers, usually comprises only a small proportion of the total populations involved. Each generation, about thirty percent of the adult male population in primitive societies perish in warfare. For modern industrial societies this figure is probably only one or two percent. Further research on this question is necessary before any conclusions can be reached.

A theory of population control in primitive society involving the effects of female infanticide and warfare has been developed by Divale (nos. 23, 187, 188, 189, 1275, 1584) and by Harris (nos. 40, 193). The theory holds that every human society must take steps to control population growth. This is especially the case in primitive societies because their simpler technology places greater limits on their ability to expand food energy production. The manner in which most primitive societies regulate their population is postulated as follows:

Infanticide is practiced on both males and females for a variety of reasons. However, since there is a general preference to have a boy as the first child, and since the ratio of males and females at birth is almost equal, many more girl infants are killed. In terms of population control this is significant because excess females are eliminated before they reproduce. The effect of selective female infanticide is that many more boys reach maturity than do girls and a shortage of marriageable women exists among young adults. The women shortage leads to adultery, rape, and wife-stealing which in turn lead to frequent disputes over women. Deaths which result from these disputes lead to blood-revenge feuding and warfare in which the excess male population is eliminated. In the childhood generation boys greatly outnumber girls because of female infanticide. But in the adult generation the ratio between the sexes is balanced because males die in warfare. However, even though the adult ratio is about equal a relative women shortage still exists due to the practice of polygyny (two or more women married to the same man). The older and more influential males have several wives, leaving younger males wifeless. The constant warfare of these societies creates a constant need for warriors and it is this need, I believe, which causes the cultural preference for a boy as a first child which begins the process in the first place.

The root of this entire system is a culturally produced women shortage. Female infanticide and polygyny create a shortage of women which leads to wars which in turn favor male infants, etc. The cycle is continuous and each generation creates conditions which perpetuate the process in succeeding generations. The next effect is the control of excess population. This theory has been successfully tested cross-culturally in the several studies cited above. An interesting aspect of this research is the light it throws on the argument about warfare and human nature. It appears that warfare is not innate in man; rather it is a learned cultural institution with several important functions. The implication is that if the conditions which require warfare are alleviated then it would be possible to do away with the institution. However, we must first learn what functions warfare serves.

This Introduction makes no attempt to discuss all the recent anthropological research on warfare. I have selected only those areas which I considered to be the most relevant and with which I was most familiar. For a broader picture the reader should consult Otterbein (no. 71) and Levine and Campbell (no. 55). The reader should also consult the Introduction to the First Edition in the pages that follow for a general description of primitive warfare which is devoid as much as possible of theoretical interpretation.

Several individuals were instrumental at various stages in the production of this bibliography; I would like to acknowledge their gracious help and to thank them sincerely. Dick Burns edited the first edition and gave me much excellent advice. It was mainly under his encouragement that I decided to compile this bibliography. Arthur Niehoff read and commented on the Introduction to the First Edition. Barbara Monahan, the production editor for the present edition, corrected several errors and I wish to thank both Barbara and the American Bibliographic Center—Clio Press for making the present edition possible. Finally, I wish to thank my assistants Mindy Chadrow and Richard Volkmann. Mindy found many of the new sources for the revised edition and Richard compiled the indexes and spent countless hours checking details. Any errors or shortcomings that remain are entirely my responsibility.

WILLIAM TULIO DIVALE
Instructor in Anthropology

York College of the City University of New York
September, 1973

INTRODUCTION TO THE FIRST EDITION

The intent of this introduction is to present a generalized description of the major aspects of primitive warfare. However, before going further two points should be clarified. First, the term "primitive" as used here refers to societies at the band, tribe, and—to some extent—the chiefdom level of cultural evolution.[1] It is cultures which are termed primitive and not their members; for no significant intellectual differences have been found between various humans throughout the world for at least the last thirty or forty thousand years. Secondly, primitive warfare is largely an aspect of the past. By the end of World War I most primitive societies had been contacted and pacified by colonial governments. Today, primitive warfare is still practiced only in the interior of Brazil and in a few isolated pockets of Oceania and South East Asia.

Yet even though primitive war is largely a thing of the past, it still has relevance. In recent years the popular writings of Robert Ardrey (AFRICAN GENESIS, no. 249, and THE TERRITORIAL IMPERATIVE, no. 231), as well as the work of Konrad Lorenz (ON AGGRESSION, no. 300), have reopened the question of whether or not war is a biological aspect of human nature. The affirmative argument usually proceeds as follows: 1) man is aggressive; 2) aggression is biologically based; 3) war is a form of aggression, therefore, war will always be a part of human society. Opponents of the aggression argument question equating aggression with warfare. They believe there is no connection

[1]Human societies in particular, and culture in general, have undergone a process of evolution. Comparing different levels of social complexity, energy-harnessing, and population density gives the following typology of development: Bands, Tribes, Chiefdoms, Stratified Societies, Pristine States, and Industrial States. For a more complete discussion of cultural evolution, see Service, 1962; Fried, 1967; White, 1959; Sahlins and Service, 1960; Harner, 1970; and Tatje and Naroll, 1970. In Anthropology, primitive warfare refers to the various types of organized violence engaged in by peoples whose societies are at the Band, Tribal, and, to a certain extent, Chiefdom levels of cultural evolution.

between the two: that one has a biological base, while the other is a cultural invention. Studies in animal behavior support the view that much of aggressive behavior is learned (see Reynolds, 1968, pp. 209-210). To claim that war makes use of individual aggression is one thing, but to claim that aggression causes war is quite another. The relevance of primitive warfare to this controversy is that it may contain insights into the origins of aggression and its cultural expressions in human society.

I

In comparison with other aspects of primitive culture, little systematic analysis has been given to war until recently: the notable classical works are Davie, no. 20; Turney-High, no. 95; and Wright, no. 108. Our failure to understand the dynamics and functions of primitive war is ironic because warfare was a common and dominant aspect of primitive culture. And yet, perhaps the reason that the importance of war in primitive societies has been overlooked is due to the cultural biases of Western observers. Members of industrial societies tend to view war as an extra-normal event or unfortunate condition of life. In addition, for reasons of trade, taxation (used to force primitives to perform wage labor for colonial companies), and administration, colonial governments quickly suppressed the warfare of the peoples under their jurisdiction. Thus quite often by the time competent observers had arrived, warfare had ceased to be practiced.

Peoples in primitive societies throughout the world lived more or less in a state of perpetual warfare. If the nature and intensity of the warfare varied greatly from people to people, it was still almost always present and an integrated part of their cultural system. As economic or territorial conquest motives were absent, except with groups at the chiefdom or higher levels of cultural evolution, the principle of blood-revenge maintained primitive warfare; but that only explains why men fight and not why warfare occurred in the first place.

The answer to the question of why primitive groups had warfare may be understood when a few important factors are considered. First, since primitive warfare occurred almost universally, it is an indication that warfare served an important function in the cultural-ecological adaptation of primitive cultures. Second, since warfare occupied an important place in each primitive society and, for most groups, pervaded almost every aspect of social and individual behavior, it must be viewed as an integral aspect of these societies. This is to say that warfare was a normal condition in primitive culture and did not represent disequilibrium.

Ecologically oriented studies conducted in the past two decades indicate that, overall, primitive societies were in ecological harmony with their environments.

It should also be stressed that primitive warfare can only be considered "normal" and contributing to equilibrium when bands fight other bands and tribes fight among tribes. While this was the case for most primitive warfare, there were instances when an expanding and more complex primitive culture would come into conflict with a less complex culture, such as a chiefdom expanding into an area inhabited by tribes. The factors that would allow a culture to increase in complexity are outside the scope of this essay, but what is relevant here is that primitive warfare in such an instance was the result of the disequilibrium that resulted from the clashing of cultures of varying complexities. This type of warfare could not go on indefinitely and would end when a new equilibrium was established, with, for example, the retreat of the tribals into a remote area that is not ecologically exploitable by the more dominant chiefdom.

The importance of seeing warfare as an integral part of primitive life is that it turns our attention to other aspects of primitive culture. And it is from this perspective of viewing warfare amidst other traits of cultural-ecological adaptation that the question of "why war?" may be answered.

It might be argued that the primary function of primitive warfare was to control the excessive male population. If the average tribesman did not see warfare in this light, nevertheless, the effects of primitive warfare interpreted from the standpoint of cultural adaptation resulted in the deaths of approximately thirty percent of the adult male generation in primitive society. Primitive warfare is more complex than this, since other cultural aspects are involved which on the surface appear to have little or no direct connection with the actual warfare to be described in this introduction. This complexity of interaction, or primitive cultural ecology, is perhaps a partial reason that a complete understanding of primitive warfare has eluded us for so long.[2]

II

Primitive war has been defined as "organized armed conflict between members of the relatively small, stateless societies traditionally studied by anthropologists" (Vayda, no. 101:48). This is a good working definition of primitive war to which we can add that the "organized armed conflict" generally falls into the categories of feuding, raiding, and open pitched battles. Quincy Wright, in an

---

[2]It is this author's position that the other cultural traits which combine with warfare to produce what can be called the "warfare syndrome" are female infanticide, which controlled the excess female population; polygyny, which worked as a built-in source of conflict as a result of an unequal distribution of women to insure a continuation of warfare; and political alliances bonded by groups exchanging women in marriage, which worked as regulatory devices in limiting the intensity of primitive warfare. For a fuller discussion of this perspective, see nos. 23, 187, 188, and 1275.

exhaustive study of war, gives a compact description of primitive warfare in his definition of "Social War":

> . . . warriors consist of all men of the tribe trained in the war moves from youth. Tactics involve little group formation or cooperation but consist of night raids, individual duels in formal pitched battles, or small headhunting or blood-revenge parties. War is initiated and ended by formalities, often quite elaborate. Its purpose is blood revenge, religious duty, individual prestige, sport, or other social objective. It may on occasion involve considerable casualties in proportion to the population of the group and is characterized as cruel or bloody by some writers because prisoners are not taken. Land or booty of economic value is not taken either. The object is slaughter of the enemy or acquisition of trophies, such as heads or scalps, of symbolic significance (Wright, no. 108:546).

Feuding is organized violence that occurs within the "group" as opposed to warfare which occurs between "groups." The term "group" varies widely according to different societies and their level of cultural evolution. In Australia, the collection of families forming the band would be the group, whereas in Amazonia the one or two segmentary lineages that form villages would be the group. Fighting within the group would be termed feuding, whereas fighting between different villages would be termed warfare. Feuding was most common in band society but existed among tribal peoples as well.

There seems to be a direct correlation between population density and the forms of violence that occurred. Bands of very low population densities, such as the Central Eskimo, practiced mostly feuding, conducted raids only infrequently and engaged in an open battle perhaps only once in a century. Among tribes which had a higher population density than bands, feuding was important, but took a second place to the raiding or war party type of warfare. Feuding was also the primary process through which groups fissioned or split apart. A feud within a village tended to split the village along lines of descent and marriage ties with the smaller faction usually moving away to found a new village. Even among the more densely populated fully tribal peoples, such as in highland New Guinea, feuds played a much smaller role than raids and battles. Among full tribals most of the fighting was with other groups. For the group to compete effectively in warfare, social solidarity and maximum group size were necessary and advantageous. Hence there was cultural selection against internal feuding and for mechanisms which maintained group unity and solidarity.

The social units that engage in feuds are generally very small. Sometimes it is simply individual activity. A typical pattern is for the offended individual to ambush or murder the man who committed the offense against him. One of the murdered man's relatives then would avenge his death by killing his murderer. This blood-revenge principle guides all primitive warfare, but in societies that raid and have battles, revenge can be satisfied simply by killing any member of the other group. Sometimes, though, if a man commits outrageous acts, his relatives will not support him in a feud, or his own kin may kill him in order to prevent a feud from developing. (Feuding was also common in peasant societies in Europe and Asia until the end of the 19th Century and is still practiced in many parts of South America; but such feuding is outside the scope of this discussion.)

The "raid" was probably the most common form of primitive warfare, and it was responsible for the highest number of deaths. The raiding or war party usually had at least ten men, but could number over a hundred, and was organized by the village headman or the men from a family with a death to avenge. Group size was very important in primitive warfare because political strategy revolved around attacking weaker and smaller groups with a lesser ability to strike back. However, there were built-in mechanisms in the social organization of primitive cultures which militated against groups becoming too large because of their limited technological, economic, and ecological resources and ability to maintain large and dense populations.

Most primitive warfare was intracultural in nature—that is, Yanomamo villages (Venezuela) raided other Yanomamo villages, and Kapauku villages (New Guinea) fought other Kapauku. Sometimes though the warfare would be along the borders of neighboring cultures, such as between the Alaskan Eskimo and the Indians of the interior. Also, if high mobility was possible because of geographical conditions, such as rivers and navigable bodies of water, or domesticated animals, such as horses, then warfare between tribes rather than between different groups of the same tribe became possible. For example, the Vikings were able to make raids along the coasts of England and Normandy, and the Plains Indians were able, with their horses, to travel the long distances necessary to raid other Plains tribes.

The war party, after leaving the village, sometimes amidst much ceremony, entered enemy territory with great caution. Surprise attack was the key element in this type of primitive warfare, and the war party usually employed an ambush. Occasionally the enemy village was attacked outright, but this was usually impossible because of various defense alarm systems, such as barking dogs and dried brush placed around the village. At dawn the raiders hid around the enemy village or camp. Upon awaking, the enemy would leave the village, individually or in small groups, and go into the bush to relieve themselves. At this point the attack was launched. The war party tried to kill as many people as possible, take the necessary heads, or scalps, whatever women they could capture, and flee, usually running for hours until it was fairly certain they were not being pursued.

A raid was considered a failure if any member of the war party was killed, regardless of how many enemy were slain. Attacking warriors would kill every male enemy they could, but they usually captured young girls and women. Quite often raids were a complete failure in the sense that no enemy were slain, or worse, only an attacker was killed. Sometimes a slaughter ensued, but most often only one or two people were killed on a raid. Casualties were considerable though because the total number killed mounted up from the sometimes dozens of these raids that occurred each year. When the war party returned home, there was often dancing and celebrating, although the warriors who made the raid usually took little part, thus giving everyone an opportunity to engage, at least indirectly, in the warfare and to achieve greater solidarity and cohesion for the group.

As with all warfare, treachery was an acceptable strategy of primitive societies. A standard procedure was for one group to invite another to its village or camp for a feast. In primitive politics intergroup or intervillage feasting was a process by which alliances were built. Even if the group invited to the feast suspected a doublecross, they would usually accept the invitation because to refuse might imply fear, which in tribal politics was an open invitation to be attacked. A third group in treacherous complicity with the host village would wait in hiding, to attack when the visitors were drunk or sleeping or about to return home. Sometimes the hosts themselves would suddenly turn on their unsuspecting guests. Treachery usually resulted in tremendous slaughter because the victims were unaware and in close physical proximity to their killers. It was not uncommon for most of the visiting men to be murdered and most of the women to be stolen.[3]

The final type of primitive warfare was the pitched battle, which involved anywhere from two hundred to two thousand warriors and was held in a predefined area or no-man's land along the borders of the warring groups. Each army was composed of warriors, usually related by marriage, from several allied villages. Even though large numbers of warriors were involved, there was little or no organized military effort; instead, dozens of individual duels were engaged in. Each warrior shouted insults at his opponent and hurled spears or fired arrows. Agility in dodging arrows was highly praised and young warriors pranced about. The women often came to watch these wars and would sing or goad their men on. Women also retrieved spent enemy arrows so that their husbands could shoot them back at their foes. Regularly occurring pitched battles were generally found among advanced tribal people in fairly dense populations. For instance, this type of warfare was not found in Amazonia, but it was common in highland New Guinea where population density is ten times that of the former area.

In spite of the huge array of warriors involved in these pitched battles, little killing took place. Because of the great distance between warriors and the relative inefficiency of primitive weapons, combined with a young warrior's

[3] A vivid account of one such treacherous feast is given by Helen Valero, a Brazilian girl who was kidnapped by the Yanomamo for twenty years (Biocca, No. 1171, pp. 194-95). Other accounts of treachery can be found in the tale of the Roman rape of the Sabine women (Plutarch:33) and in the story of Jacob's sons who avenged the rape of their sister Dinah (Genesis 34:25-31).

agility to dodge arrows, direct hits rarely occurred. In the event that someone was badly wounded or slain, the battle would usually cease for that day. Groups that fought in pitched battles also conducted raids or ambushes, and it was here that most of the killing occurred. In the past, many anthropologists viewed these pitched battles and, noting the small number of casualties, concluded that much or all of primitive warfare was a ritual or game. However, this perspective is now questioned, and it is suggested that such warfare was extremely effective, perhaps even overeffective, in the sense that many cultural controls existed whose primary aim was the regulation and limitation of warfare.

III

The purpose of primitive warfare was revenge against an earlier injustice or killing perpetrated by the enemy, and the sole object was to kill the enemy. If weapons or food were come upon in the process of attacking, these would be taken. Young girls and women were also stolen, but infants, boys, and men were always slain. Some groups, especially those practicing cannibalism, would take war captives. This practice has sometimes been mistaken for slavery, but there was no economic intention involved. Even though the stealing of women and food was common and considered a boon to the war party, a raid would be initiated only for revenge. Some exceptions to this rule existed among pastoralists who, in addition to revenge, would also raid to steal their enemies' herd animals. For example, with the advent of horses among the Plains Indians, raiding for horses began. Among bands or tribes, however, warfare was rarely conducted to conquer additional territory as a result of population pressures on land.[4] Sometimes groups would go to war over encroachments made on their hunting territory or horticultural gardens, but these disputes, which did not occur very often, were more concerned with use-rights of certain natural resources rather than with territory.

Among groups at the chiefdom level of cultural evolution, warfare over land started to become a factor which operated along with warfare conducted for blood-revenge. During the chiefdom stage in Mesopotamia, for example, most of the wars between the various chiefdom centers were over strips of land between their borders (Divale, no. 1371). Vayda also describes the territorial conquest aspects of warfare among the Maori, who were incipient chiefdoms and in general terms this is also discussed by Otterbein (nos. 69, 240).

Primitive wars usually began when someone was killed and the dead man's relatives, under the obligation of the blood-revenge principle, attempted to

---

[4] As mentioned earlier (see footnote no. 1), the adaptive function of primitive warfare was to relieve pressure from population growth in band and tribal society.

retaliate and even the score. Women were the most frequent cause of these disputes that led to killings. The most common arguments were charges of adultery involving the irate husband and his wife's lover, but equally as common were disputes over brideprice (the wealth given to the bride's relatives by the relatives of the groom). Divorce also often caused wars. If a woman wanted a divorce or if she ran off with another man, her husband and his relatives would want a return of the brideprice. The bride's family might refuse, and in the event that the groom felt cheated, this constituted a reason to go to war. Among the Kapauku of New Guinea, for example, almost half of the wars stemmed from this cause.

Among most groups, however, it was adultery and wife-stealing that caused most wars. If a husband caught his wife in an adulterous affair, he might brutally beat or even disfigure her and attempt to kill her lover. If he was successful, a new war would begin. If her lover was from the husband's village, however, some attempt might be made to keep the peace by the lover's family giving the husband economic retribution. Often, though, the husband and his wife's lover would fight, causing the village to split. This fissioning process was common all over the world and appears to be the primary way in which new villages were generated. Wife-stealing would trigger wars when one man talked another man's wife into running off with him or when women of other villages were stolen outright. To a hunting or war party that accidentally came across women gathering food or firewood, temptation might prove too great; thus women were taken as by-products of successful raids, but it was rare for a raid to be conducted solely to steal women.

Primitive wars had other causes besides disputes over women, but they formed a minority of cases. Some disputes resulted from one group infringing upon the hunting or gathering territory of another without getting prior permission. Among horticulturalists the infringement may take the form of planting a garden in another group's area or of stealing food from another group's garden. In New Guinea, if the pigs of one man damage the garden or property of another, a war can result should the owner of the pig refuse to pay for the damages.

Several anthropologists report that charges of witchcraft and sorcery were also the cause of many primitive wars. This notion, however, probably confuses cause and effect. As many primitive societies do not believe that death can occur from natural causes (such as sickness, disease or accidents), whenever a death of this type occurs it is usually charged to witchcraft or sorcery. The kinsmen of the dead person naturally ask "who was the witch?" and the response, almost invariably, is an enemy of the dead man. The relevant point is that charges of witchcraft and sorcery were usually directed against individuals or groups where prior disputes were present. It is suggested that witchcraft be viewed, in respect to warfare, as a mechanism for maintaining group solidarity and hate for the enemy, rather than as a cause of primitive warfare.

This discussion has primarily described primitive warfare and has shown how it was part of the everyday social reality of primitive society and that, depending on the population density of the groups involved, certain types of fighting predominated, such as feuding, raiding, or formal battles. Coordinated activity or battle strategy was almost nonexistent. Treachery and the hit-and-run ambush were the most common and most effective modes of fighting. Primitive warfare was conducted for purposes of blood-revenge and not for economic motives. Finally, disputes over women were the major cause of primitive wars.

WILLIAM TULIO DIVALE

Los Angeles, California
August, 1971

REFERENCES CITED

(which are not included in this Bibliography)

Fried, Morton. *The Evolution of Political Society.* New York: Random House, 1967.

Harner, Michael. "Population Pressure and the Social Evolution of Agriculturalists." *Southwestern Journal of Anthropology,* (1970), 26:67-86.

Murdock, George P. *Outline of World Cultures.* 3rd Revised Edition. New Haven: Human Relations Area Files, 1963.

———. *Outline of Cultural Materials.* 4th Revised Edition. New Haven: Human Relations Area Files, 1961.

———, et al. "Ethnographic Atlas." *Ethnology,* (1962-7), 1 (1)-6 (1).

Reynolds, V. "Kinship and the Family in Monkeys, Apes and Man." *Man,* (1968), 3:209-23.

Sahlins, M., and E. Service, eds. *Evolution and Culture.* Ann Arbor: University of Michigan Press, 1960.

Service, E. *Primitive Social Organization.* New York: Random House, 1962.

Tatje, T., and R. Naroll. "Two Measures of Societal Complexity: An Empirical Cross-cultural Comparison." In *A Handbook of Method in Cultural Anthropology,* eds. Raoul Naroll and Ronald Cohen, pp. 766-833. New York: Natural History Press, 1970.

White, Leslie. *The Evolution of Culture.* New York: McGraw Hill, 1959.

A NOTE ON PRIMITIVE WAR/PEACE RESEARCH

Most of the literature on primitive warfare dealing with specific societies is not presented as such bibliographically, but is included as a chapter or a few pages in general ethnographic accounts of primitive societies. An ethnography is a descriptive and analytical account of a culture written by an observer, usually an anthropologist, who has lived among the people he is describing. Since warfare was the first aspect of primitive society suppressed by colonial governments upon making contact with these peoples, and missionaries—who usually followed on the heels of the colonial police—worked vigorously to eradicate the war mores of primitive peoples, it is quite common for ethnographic accounts (sometimes written by anthropologists many years after pacification) to give only a passing mention to warfare, or for the discussion to be simply an historical reconstruction. Thus, the researcher will find it worthwhile to study the early accounts of primitive peoples in the written journals of explorers, travelers, missionaries, and colonial administrators.

The researcher who wishes to study the warfare of particular cultures or regions of the world should use this bibliography as a stepping stone from which to investigate the general ethnographic literature on the societies of his interest. A good source for locating the various cultures in the world is the *Outline of World Cultures* (Murdock, 1963) where societies are listed by geographical location and indexed by name. Bibliographical information on 1,170 of these societies can be found in the *Ethnographical Atlas* (Murdock, et. al., 1962-7).

An excellent tool for use in cross-cultural or comparative research on warfare is the Human Relations Area Files (HRAF), which are located in either the original or microfile form at many large university libraries. Information on 290 different cultures of the world is collected and coded into 88 general topics in these files (see Murdock, et al., 1961 for specific coding procedures.) In the HRAF, file category no. 72 is War which is further divided into nine subcategories with such topics as "instigation of war," and "peace making."

USE OF THIS BIBLIOGRAPHY

This bibliography is divided into topical and geographical sections. Part I is divided into sixteen topical categories (see Table of Contents). For the most part, the sources listed here are of a theoretical nature, or they concern a specific topic such as headhunting. These topical categories are fairly well defined and no category is too large for the researcher to examine every title included in it. Part II is divided into seven major geographical regions. The sources listed here refer to the warfare of the various peoples of these regions. General works cited in this section are followed by their page numbers which contain descriptions or discussions of primitive warfare. Some of the sources in the topical section refer to various tribal peoples in the geographical section. For example, an article on scalping practices of the Sioux is listed under topic. M., "Scalping and War Trophies." But references to warfare among the Sioux are listed in the geographical section under Q.3, "The Great Plains of North America." To aid the researcher who is interested in all aspects of warfare among a particular culture and to compensate for the overlap of the topical and geographical categories, the following procedure was adopted: An index is provided listing alphabetically all of the cultures referred to in this bibliography. Each source is numbered and those which pertain to a culture are listed after that culture's name as it appears in the index.

For the reader who is not interested in using this bibliography for research purposes, but instead wishes to have a general knowledge of primitive warfare and what anthropologists have concluded about it, a selection of twenty-five sources has been made. This selection contains both the classic works on this topic as well as the results of the latest research (see the following source numbers: 8, 15, 20, 23, 33, 52, 55, 56, 64, 69, 71, 92, 95, 97, 98, 99, 107, 109, 127, 193, 204, 239, 245, 390, 1179).

ABBREVIATIONS OF PERIODICALS & ANTHOLOGIES CITED

AA   *American Anthropologist*
AAOJ   *American Antiquarian and Oriental Journal*
AAR   *Annual Archaeological Report* (See appendix to *Report of the Minister of Education,* Ontario)
AHQ   *Alabama Historical Quarterly*
AI   *American Indigena*
AJPA   *American Journal of Physical Anthropology*
AJS   *American Journal of Sociology*
AJSL   *American Journal of Semitic Languages and Literatures,* Chicago
AMJ   *American Museum Journal*
AN   *American Naturalist*
APAM   *Anthropological Papers of the American Museum of Natural History*
AR   *Asiatic Review*
ARBAE   *Annual Reports of the Bureau of American Ethnology*
ARSI   *Annual Report of the Board of Regents of the Smithsonian Institution*
ASR   *American Sociological Review*
BASOR   *Bulletin of the American Schools of Oriental Research,* New Haven
BBAE   *Bulletins of the Bureau of American Ethnology*
BBSNS   *Bulletin of the Buffalo Society of Natural Sciences*
BJS   *British Journal of Sociology*
BTLV   *Bijdragen tot de Taal-Land en Volkenkunde*
CES   *Comparative Ethnographic Studies*
CHC   *Coleccion de Historiadores de Chile*
CLD   *Coleccion de Libros y Documentos Referentes a la Historia de America,* Madrid
CMHS   *Collections of the Minnesota Historical Society*
CO   *Chronicles of Oklahoma*

CTCT   *Conflict Tension and Cultural Trend in India.* L.P. Vidyarthi, ed. Calcutta: Punthi Pustak, 1969.

EP   *El Palacio*

FHSQ   *Florida Historical (Society) Quarterly*

FMSW   *Franciscan Missions of the Southwest*

FTD   *For the Dean.* Essays in Anthropology in Honor of Byron Cummings. Tucson, 1950.

HAHR   *Hispanic American Historical Review*

HSAI   *Handbook of South American Indians.* Julian H. Steward, ed. 7 vols. *Bulletin of the Bureau of American Ethnology, no. 143.* Washington: Government Printing Office, 1946/59.

HSS   Works Issued by the Hakluyt Society, Second Series, London

ICA   *Proceedings of the International Congress of Americanists*

IEJ   *Israel Exploration Journal*

IH   *Indian Historian*

IN   *Indian Notes,* Museum of the American Indian, Heye Foundation, New York

INM   *Indian Notes and Monographs,* Museum of the American Indian, Heye Foundation, New York

IS   *Inland South America* (Our Flag and Field, S. American Indians)

JAF   *Journal of American Folklore*

JAFL   *Journal of American Folklore*

JAI   *Journal of the Royal Anthropological Institute of Great Britain and Ireland*

JASB   *Journal of the Anthropological Society of Bombay*

JCR   *Journal of Conflict Resolution*

JEA   *Journal of Egyptian Archaeology*

JLPS   *Journal of Legal and Political Sociology*

JNES   *Journal of Near Eastern Studies*

JPS   *Journal of the Polynesian Society*

JSAP   *Journal de la Societe des Americanistes,* Paris

LWSAC   *Law and Warfare: Studies in the Anthropology of Conflict.* Paul Bohannan, ed. New York: Natural History Press, 1967.

M   *Masterkey*

MAAA   *Memoirs of the American Anthropological Association*

MAES   *Monographs of the American Ethnological Society*

MAGW   *Mitteilungen der Anthropologischen Gesellschaft in Wien*

MHSC   Collection of Massachusetts Historical Society, Cambridge

MNMNA   *Museum Notes of the Museum of Northern Arizona*

NH   *Natural History*

NHA   *The Natural History of Aggression.* J.D. Carthy and F.J. Ebling, eds. New York and London: The Academic Press, 1964.

NHC   *The Nature of Human Conflict.* Elton B. McNeil, ed. Englewood Cliffs, New Jersey: Prentice-Hall, 1965.

NMHR   *New Mexico Historical Review*
    P   *Plateau*
PAAAS   *Proceedings of the American Association for the Advancement of Science*
 PAPS   *Proceedings of the American Philosophical Society*
  PEQ   *Palestine Exploration Quarterly*
   PH   *Primitive Heritage.* Margaret Mead and Nicolai Calas, eds. New York, 1953.
  PIA   *Plains Indians Anthropologists*
 PKAS   *Publications of the Kroeber Anthropological Society*
 PPHR   *Panhandle-Plains Historical Review*
 PPSC   *Proceedings of the (fifth) Pacific Science Congress*
PSHSW   *Proceedings of the State Historical Society of Wisconsin*
PTRSC   *Proceedings and Transactions of the Royal Society of Canada*
RACHS   *Records of the American Catholic Historical Society of Philadelphia*
RCAETS *Reports of the Cambridge Anthropological Expedition to Torres Straits.* A.C. Hadon and A. Wilkin. 1904.
 RUNC   *Revista de la Universidad Nacional de Cordoba*
RUSNM   *Reports of the United States National Museum*
 SPVW   *Science and Psychoanalysis. vol. 6: Violence and War.* Jules Masserman, ed. New York: Grune and Stratton, 1963.
   SW   *Southern Workmen*
SWAA-   Papers presented at the 1971 Joint Annual Meeting of the
  AES   Southwestern Anthropological Association and the American Ethnological Society (April 29-May 1, Tucson, Arizona)
 SWHQ   *Southwestern Historical Quarterly*
 SWJA   *Southwestern Journal of Anthropology*
 TIJD   *Tijdschrift voor Indische Taal-, Land-en Volkenkunde,* Batavia
  TNR   *Tanganyika Notes and Records*
 TNYS   *Transactions of the New York Academy of Sciences*
TSHAQ   *Texas State Historical Association Quarterly*
  UHQ   *Utah Historical Quarterly*
  VSE   *Violence and the Struggle for Existence.* David N. Daniels, M.F. Gilula, and F.M. Grenberg, eds. Boston: Little, Brown, 1970.
WAACA   *War: The Anthropology of Armed Conflict and Aggression.* Morton Fried, Marvin Harris, and Robert Murphy, eds. New York: Natural History Press, 1968.
  WHR   *War and the Human Race.* Maurice N. Walsh, ed. New York, Amsterdam: Elsevier Publishing, 1971.
YPMCM   *Yearbook of the Public Museum of the City of Milwaukee*
 YUPA   *Yale University Publications in Anthropology*
  ZFE   *Zeitschrift fur Anthropologie, Ethnologie und Urgeschichte*

Part I

Topical

A / PRIMITIVE WARFARE: GENERAL

1 "Ancient World of a War-torn Tribe." *Life*, (1968), 53:73-91.

2 ANDRESKI, Stanislav. "Origins of War." NHA, (1964), 129-136.

3 ARON, Raymond. *On War*. Garden City, 1959.

4 BASU, M.N. "Conflicts and Tensions in Human Society." CTCT, (1969), 1-14.

5 BEALS, A.R., and E.B. Siegel. *Divisiveness and Social Conflict: An Anthropological Approach*. Palo Alto, California: Stanford University Press, 1966.

6 BENEDICT, Ruth. "The Natural History of War." In M. Mead, *An Anthropologist at Work*. Boston: Houghton Mifflin, 1959, pp. 369-82.

7 BERNARD, J., T.H. Pear, R. Aron, and R.C. Angell, eds. *The Nature of Conflict*. Paris: UNESCO, 1957.

8 BOHANNAN, Paul, ed. *Law and Warfare: Studies in the Anthropology of Conflict*. New York: Natural History Press, 1967.

9 BOUTHOUL, Gaston. *Les Guerres: Elements de polemologie*. Paris, 1951.

10 BRAMSON, L., and G.W. Goethals, eds. *War: Studies from Psychology, Sociology, and Anthropology*. New York: Basic Books, 1964.

11 BRODIE, B. "Theories on the Causes of War." WHR, (1971), 1-11.

12 BROWN, Paula. "Enemies and Affines." *Ethnology*, (1964), 3:335-56.

13 CAMPBELL, Donald, and Robert Levine. "A Proposal for Cross-cultural Research on Ethnocentrism." JCR, (1961), 5:82-108.

14 CAMPBELL-SMITH, M. "Enemy." *Hasting's Encyclopedia of Religion and Ethics*, v. 5. Edinburgh (1912).

3

15 CARNEIRO, Robert L. "A Theory of the Origin of the State." *Science*, (1970), 169:733-38.

16 CATTON, William R., Jr. "The Functions and Dysfunctions of Ethnocentrism: A Theory." *Social Problems*, (1961), 8:201-211.

17 CHILDE, V.G. "War in Prehistoric Societies." *Sociological Review*, (1941), 33:126-38.

18 CLAUSEWITZ, Karl von. *On War*. F.N. Maude, ed., J. Graham, trans. Revised edition, 3 volumes. New York: Barnes & Noble, Inc., 1961.

19 COHEN, John. "Human Nature, War and Society." *The Thinkers Library*, no. 112. London, (1946).

20 DAVIE, Maurice R. *The Evolution of War: A Study of Its Role in Early Societies.* Port Washington, New York: Kennikat Press, 1968 (original Yale University Press, 1929).

21 DENTON, Frank H., and W. Phillips. "Some Patterns in the History of Violence." JCR, (1968), 12:182-95.

22 DIAZ DE VILLIZGAS, B.J. "Los origenes remotos de la guerra, Las primeras locallas de qua hay Noticia Se Libranon en Espana." *Boletin Arqueologico de Sudeste Espanol*, (1949), 12-15:111-20.

23 DIVALE, William T. "Systemic Population Control in the Middle and Upper Paleolithic: Inferences Based on Contemporary Hunter-gatherers." *World Archaeology*, (1972), 4:222-243.

24 ——. "Migration, External Warfare, and Matrilocal Residence: An Explanation for Matrilineal Residence Systems." *Behavior Science Notes* (1973) 8:00-00, in press.

25 ——. *The Causes of Matrilocal Residence: A Cross-Ethnohistorical Study.* Ph.D. thesis, Dept. of Anthropology, State University of New York at Buffalo, 1973.

26 DOWNS, James F. "Thoughts on Cavalry, Guerilla Warfare and the Fall of Empires." *Kroeber Anthropological Society Papers*, (1960), 23:105-13.

27 DYK, Walter. *A Study of the Effect of Change of Technique on the Warfare of Primitive Peoples.* M.A. thesis, University of Chicago, 1931.

28 EMBER, Melvin and Carol. "The Conditions Favoring Matrilocal versus Patrilocal Residence." AA, (1971), 73:571-594.

29 FALLS, Cyril. *The Place of War in History*. Oxford, 1947.

30 FERRERO, Guglielmo. *Militarism, and Contribution to the Peace Crusade*. Boston: L-C. Page, 1903.

31 FLANIGAN, William H., and E. Fogelman. "Patterns of Political Violence in Comparative Historical Perspective." *Comparative Politics*, (1970), 3:1-20.

32 FORTUNE, R.F. "The Rules of Relationship Behavior in One Variety of Primitive Warfare." *Man*, (1947), 47:108-110.

33 FOX-PITT-RIVERS, A.H.L. "Primitive Warfare." In *The Evolution of Culture*. Oxford: Clarendon Press, 1906, pp. 45-185.

34 FRIED, M.H. "Review of Bramson, 1964, 'War . . .' " AA, (1965), 67:1344-46.

35 FRIED, Morton, Marvin Harris, and Robert Murphy, eds. *War: The Anthropology of Armed Conflict and Aggression*. New York: Natural History Press, 1968.

36 FURER-HAIMENDORF, Christoph von. "Review of *War: The Anthropology of Armed Conflict and Aggression*, M. Fried, M. Harris, and R. Murphy, eds." *Saturday Review*, (1968), 51:27-29.

37 GINSBURG, Morris. "The Causes of War." *Sociological Review*, (1939), 31:121-43.

38 HAMBLY, Wilfrid, D. "How Primitive Man Waged War." *Travel*, (1947), 88:18, 21, 31.

39 ———. "Primitive Warfare: So-called 'Savages' Fight Less, and Less Cruelly, than 'Civilized' Nations." *Chicago Natural History Museum Bulletin*, (1946), 17(9-10):4-5.

40 HARRIS, Marvin. "Warfare Old and New." NH, (1972), 81(3):18, 20.

41 HOBBES, Thomas. *Leviathan*. Oxford: Clarendon Press, 1943 edition (original 1651).

42 HOIJER, Harry. "The Causes of Primitive Warfare." M.A. thesis, University of Chicago, 1929.

43 HOLLOWAY, R. "Review of E. McNeil, 1965, *The Nature of Human Conflict*." AA, (1966), 68:830-32.

44 HOLSTI, Ole R., and Robert C. North. "The History of Human Conflict." NHC, (1965), 155-71.

45 HOLSTI, R. *Relation of War to the Origin of the State*. Helsingfors, 1913.

46 KEITH, Arthur. *Essays on Human Evolution.* London: Watts, 1946.

47 KENNEDY, J. "Ritual and Intergroup Murder: Comments on War, Primitive and Modern." WHR, (1971), 40-61.

48 KING, J.C. "The Role of Warfare in History." WHR, (1971), 62-9.

49 LaGORGETTE, Jean. *Le Role de la guerre.* Paris, 1906.

50 LEA, Henry C. "The Wager of Battle." LWSAC, (1967), 233-54.

51 LEEDS, A. *Selected References on Warfare.* Dept. of Anthropology, University of Texas, mimeograph, 1967.

52 LEEDS, Anthony. "The Functions of War." SPVW, (1963), 69-82.

53 LETOURNEAU, Charles J.M. *La Guerre dans les diverses races humaines.* Paris: L. Battaille, 1895.

54 LEVINE, R. "Anthropology and the Study of Conflict: An Introduction." JCR, (1961), 5:3-15.

55 LEVINE, Robert A., and Donald T. Campbell. *Ethnocentrism: Theories of Conflict, Ethnic Attitudes, and Group Behavior.* New York: John Wiley, 1972.

56 MALINOWSKI, B. "An Anthropological Analysis of War." AJS, (1941), 46:521-50.

57 ———. "War . . . Past, Present and Future." In J. Clark and T. Cochran, eds., *War as a Social Institution.* New York, 1941.

58 ———. "Un analisis antropologico de la guerra." *Revista Mexicana de Sociologia,* (1941), 3:119-49.

59 ———. "Una analise antropologica da guerra." *Sociologia,* (1941), 3:203-17.

60 McNEIL, Elton B., ed. *The Nature of Human Conflict.* Englewood Cliffs, New Jersey: Prentice-Hall, 1965.

61 MEAD, Margaret, and Rhoda Metraux. "The Anthropology of Human Conflict." NHC, (1965), 116-138.

62 MEGGITT, Mervyn. "System and Subsystem: The Te Exchange Cycle Among the Mae Enga." *Human Ecology,* (1972), 1:111-123.

63 MONTAGU, M. Francis A. "The Nature of War and the Myth of Nature." *Scientific Monthly,* (1942), 54:342-53.

64 NAROLL, Raoul, and W.T. Divale. "Natural Selection in Cultural Evolution: Warfare versus Peaceful Diffusion." *Behavior Science Notes,* in press, (1974).

65 NEF, John U. *War and Human Progress. An Essay on the Rise of Indian Civilization.* Cambridge, 1952.

66 NEWCOMB, W.W. "Toward an Understanding of War." In Gertrude E. Dole, and R. Carneiro, eds., *Essays in the Science of Culture in Honor of Leslie A. White.* New York: Thomas Crowell, 1960, pp. 317-36.

67 NOBERINI, M. *Ethnocentrism and Feuding: A Cross-Cultural Study.* M.A. thesis, University of Chicago, 1966.

68 NOVILOW, J. *War and Its Alleged Benefits.* New York, 1911.

69 OTTERBEIN, Keith. *The Evolution of War: A Cross-Cultural Study.* New Haven: Human Relations Area Files Press, 1970.

70 ———. "Internal War: A Cross-Cultural Study." AA, (1968), 70:277-89.

71 ———. "Anthropology of War." In J.J. Honigmann, ed., *Handbook of Social and Cultural Anthropology.* New York: Rand McNally, 1973.

72 PATRICK, G. "War as a Form of Relaxation." In R.E. Park, *Introduction to the Science of Sociology.* Chicago, 1921, pp. 598-600.

73 PERRY, W.J. "An Ethnological Study of Warfare." *Proceedings of the Manchester Literary and Philosophical Society,* (1917), 61(5).

74 ———. "An Ethnological Study of Warfare." *Memoirs and Proceedings of the Manchester Literary and Philosophical Society,* (1917), 61(6).

75 ———. "Pugnacity." *The Monist,* (1923), 33:116-138.

76 PRESTON, Richard A., Sidney F. Wise, and Herman O. Werner. *Men in Arms: A History of Warfare and Its Interrelationships with Western Society.* London: Atlantic Press, 1956.

77 PRUITT, Dean G., and R. Snyder, eds. *Theory and Research on the Causes of War.* Englewood Cliffs: Prentice-Hall, 1969.

78 QUINTERO, R. "La guerra no es can antigua como el hombre." *Caracas Cultura Universitaria,* (1963), 82:7-19.

79 RAPOPORT, Anatol. *Fights, Games and Debates.* Ann Arbor: University of Michigan Press, 1960.

80 READ, G.H. "Anthropology and War." JAI, (1919), 49:12-19.

81 REINACH, A. "Les Trophees et les origines religieuses de la guerre." *Revue d'ethnographie et de Sociologie,* (1913).

82 RICHARDSON, Lewis F. *Arms and Insecurity. A Mathematical Study of the Causes and Origins of War.* eds., N. Rashevsky and E. Trucco. Pittsburgh: Boxwood Press, 1960.

83 ——. *Statistics of Deadly Quarrels.* eds., Q. Wright and C. Lienan. Pittsburgh: Boxwood Press, 1960.

84 RIGGS, A.S. "The Evolution of War." *Scientific Monthly,* (1942), 54:110-24.

85 ROSE, Arnold M. "The Comparative Study of Intergroup Conflict." *Sociological Quarterly,* (1960), 1:57-66.

86 ROSENAU, James N. "Behavioral Science, Behavioral Scientists, and the Study of International Phenomena: A Review of Leon Bramson, and George Goethals, eds., *War: Studies from Psychology, Sociology, and Anthropology.*" JCR, (1965), 9:509-20.

87 SCHELLING, Thomas. *The Strategy of Conflict.* Cambridge: Harvard University Press, 1960.

88 SCHNEIDER, Joseph. "On the Beginnings of Warfare." *Social Forces,* (1952), 31:68-74.

89 ——. "Primitive Warfare: A Methodological Note." ASR, (1950), 15:772-77.

90 SERVICE, Elman. "War and Our Contemporary Ancestors." WAACA, (1968), 160-67.

91 SIMMEL, George. *Conflict and the Web of Group Affiliations.* Glencoe, 1955.

92 SIPES, Richard G. "War, Sports and Aggression: An Empirical Test of Two Rival Theories." AA, (1973), 75:64-86.

93 SPEIER, H. "The Social Types of War." AJS, (1941), 46:445-54.

94 SWANTON, John R. *Are Wars Inevitable?* Smithsonian Institution War Background Studies No. 12, 1943.

95 TURNEY-HIGH, H.H. *Primitive War.* Columbia: University of South Carolina Press, 1949.

96 ——. *The Practice of Primitive War.* University of Montana Publications in the Social Sciences, No. 2. Missoula, Montana, 1942.

97 VAN VELZEN, H.U.E. Thoden, and W. van Wetering. "Residence, Power Groups and Intra-societal Aggression: An Inquiry into the Conditions Leading to Peacefulness within Non-Stratified Societies." *International Archives of Ethnography,* (1960), 49:169-200.

98 VAYDA, Andrew P. "Phases of the Process of War and Peace among the Marings of New Guinea." *Oceania,* (1971), 42:1-24.

99 ——. "Hypotheses about Functions of War." WAACA, (1968), 85-91.

100 ——. "War: Primitive Warfare." *International Encyclopedia of the Social Sciences*, (1968), 16:468-72.

101 ——. "Hypotheses About Functions of War." NH, (1967), 76:10, 48-50, 69.

102 ——. *Research on the Functions of Primitive War*. Peace Research Society International Papers, v. 7, 1967.

103 ——. *Selected References on Warfare*. Dept. of Anthropology, Columbia University, mimeograph, 1961.

104 ——, and A. Leeds. "Anthropology and the Study of War." *Anthropologica*, (1961), 3:131-34.

105 WALSH, Maurice N., ed. *War and the Human Race*. New York: Elsevier Publishing Co., 1971.

106 WIRSING, Rolf. "Political Power and Information: A Cross-Cultural Study." AA, (1973), 75:153-170.

107 WRIGHT, Quincy. "War: The Study of War." *International Encyclopedia of the Social Sciences*. (1968), v. 16, pp. 453-68.

108 ——. *A Study of War*. Chicago: University of Chicago Press, 1942.

109 ——. "Primitive Warfare." In Q. Wright, *A Study of War*. Chicago: Chicago University Press, 1942, pp. 53-88.

A.1 / Revolution

110 ADAMS, M.N. "The Sioux Outbreak in the Year 1862." CMHS, (1901), IX:431-52.

111 ALMOND, Nina, and N.H. Fisher. *Special Collections in the Hoover Library on War, Revolution, and Peace*. Stanford: Stanford University, 1940.

112 BOARDMAN, E.P. "Millenary Aspects of the Taiping Rebellion (1851-64)." In S.L. Thrupp, ed., *Millennial Dreams in Action: Studies in Revolutionary Religious Movements*.

113 COHEN, R. "Review of M. Gluckman, *Order and Rebellion in Tribal Africa*." AA, (1965), 67:950.

114 DIXON, G. Aubrey, and Otto Heilbrun. *Communist Guerrilla Warfare*. London, 1954.

115 EWING, C.R. "Investigations into the Causes of the Pima Uprising of 1751." *Mid-America*, (1941), XXIII:139-51.

116 FRUCHT, Richard. "Rebellion in the Caribbean: St. Kitts, Anguilla, and the 'Maffia.'" Paper presented at the 68th annual meeting of the American Anthropological Association, New Orleans, Louisiana, 1969.

117 GARCIA, L. "Les Mouvements de resistance au Dahomey (1914-1917)." *Cahiers d'Etudes Africaines*, (1970), 10:144-78.

118 GERLACH, L.P., and V.H. Hine. "The Social Organization of a Movement of Revolutionary Change: Case Study, Black Power." In N. Witten and J. Szwed, eds., *Afro-American Anthropology*. New York: Free Press, 1970.

119 GLUCKMAN, Max. *Order and Rebellion in Tribal Africa*. New York: Free Press, 1963.

120 ——. "Rituals of Rebellion in South-East Africa." In M. Gluckman, *Order and Rebellion in Tribal Africa*. New York: Free Press, 1963.

121 GRIFFITH, Samuel B. *Mao Tse-Tung on Guerrilla Warfare*. New York: Praeger, 1961.

122 ——. *Sun Tzu. The Art of War*. New York: Oxford University Press, 1963.

123 ——. *Custom and Conflict in Africa*. Oxford: Blackwell, 1955.

124 GUEVARA, Che. *Guerrilla Warfare*. New York, 1961.

125 GUEVARA, Silva T. *Los Arancanos en la revolucion de la independencia 1810-1827*. Santiago, 1910.

126 HACKETT, C.W. "The Revolt of the Pueblo Indians of New Mexico in 1680." TSHAQ, (1911), XV:93-147.

127 HOBSBAWN, E.S. *Social Bandits and Primitive Rebels: Studies in Archaic Forms of Social Movements in the 19th and 20th Centuries*. Glencoe, Illinois, 1960.

128 HOLTERMAN, J. "The Revolt of Yozcolo: Indian Warrior in the Fight for Freedom." IH, (1970), 3:19-23.

129 ——. "The Revolt of Estanislao." IH, (1970), 3:43-54, 66.

130 KAHIN, George McT. *Nationalism and Revolution in Indonesia*. Ithaca, New York: Cornell University Press, 1952.

131 LEWIN, B. *La Rebelion de Tupac Amaru*. Buenos Aires, 1957.

132 McKIBBEN, D.D. "Revolt of the Navajo, 1913." NMHR, (1954), XXXIX:259-89.

133 MEANS, P.A. "The Rebellion of Tupac-Amaru, 1780-1781." HAHR, (1919), 11:1-25.

134 MOUNTENEY-JEPHSON, Arthur J. *Emin Padcha and the Rebellion at the Equator*. London: S. Low, 1890.

135 NKRUMAH, Kwame. *Handbook of Revolutionary Warfare*. New York: International Publishers, 1970.

136 PARET, Peter, and John Shy. *Theory of Guerrilla Warfare*. New York, 1961.

137 PARKMAN, F. *The Conspiracy of Pontiac and the Indian War after the Conquest of Canada*. 2 vols. Boston: Little, Brown, 1886.

138 PECKHAM, H.H. *Pontiac and the Indian Uprising.* Princeton: Princeton University Press, 1947.

139 RITZENTHALER, R.E. "Anlu: A Women's Uprising in the British Cameroons." *African Studies* (Johannesburg), (1960), 19:151-56.

140 SWEET, G.W. "Incidents of the Threatened Outbreak of Hole-in-the-Day." CMHS, (1894), VI:401-8.

141 THRUPP, Sylvia, ed. *Millennial Dreams in Action: Studies in Revolutionary Religious Movements*. New York: Schocken Books, 1970.

142 Van Der KROEF, J.M. "Messianic Movements in the Celebes, Sumatra, and Borneo." In S.L. Thrupp, ed. (see no. 141), pp. 80-121.

143 WALLACE, Anthony F.C. "Revitalization, Violence, and Revolution." Paper presented at the 68th annual meeting of the American Anthropological Association, New Orleans, Louisiana, 1969.

144 WISSLER, Clark. "Depression and Revolt. The Story of the Last Indian Uprising and its Youth Movement." NH, (1938), 41:108-12.

145 WOLF, Eric. *Peasant Wars of the Twentieth Century*. New York: Harper and Row, 1969.

146 ZIDE, N.H., and R.D. Munda. "Revolutionary Birsa and the Songs Related to Him." *Journal of Social Research*, (1969), 12:37-60.

B / ANTHROPOLOGISTS ON MODERN WARFARE

147 "Indians as Code Transmitters." *Masterkey*, (1941), 15:240.

148 BITTKER, Thomas E. "The Choice of Collective Violence in Intergroup Conflict." VSE, (1970).

149 BOAS, Franz. "An Anthropologist's View of War." *International Concilia-tion*, (1912), (52).

150 COLLINS, H.B., A.H. Clark, and E.H. Walker. "The Aleutian Islands." *Smithsonian Institution War Background Series*, (1945), XXI:1-131.

151 DANIELS, David N., M.F. Gilula, and F.M. Gehberg, eds. *Violence and the Struggle for Existence.* Boston: Little, Brown, 1970.

152 DIAMOND, Stanley. "War and the Dissociated Personality." WAACA, (1968), 183-88.

153 DONAGHUE, John D. "Social Structure and Guerrilla Warfare in Viet Nam." Paper presented at the 62nd annual meeting of the American Anthropological Association, San Francisco, California, 1963.

154 ELKIN, Adolphus P. "Anthropology and the War." *Mankind*, (1942), 3:77-79.

155 FORTES, M. "The Impact of the War on British West Africa." *International Affairs*, (1945), 21:206-19.

156 FRIED, Morton, Marvin Harris, and Robert Murphy, eds., "Fink-out or Teach-in?" WAACA, (1968), ix-xix.

157 GILULA, M.F., and D.N. Daniels. "Violence and Man's Struggle to Adapt." *Science*, (1969), 164:396.

158 HOEBEL, E. Adamson. "The Draft and the United States Congress." WAACA, (1968), 208-10.

159 HRDLICKA, Ales. "Suggestions Relating to the New National Army by the Anthropology Committee of the National Research Council." *Proceedings of the National Academy of Sciences*, (1971), 3:526-28.

160 KAIKINI, V. "Anthropological Observations on Indian Troops during the World War." *Journal of the Anthropological Society of Bombay*, (1953), 7:45-60.

161 KEISER, R.L. "The World View of Street Warriors." In James P. Spradley, and D. McCurdy, eds., *Conformity and Conflict.* Boston: Little, Brown, 1971.

162 KEITH, Arthur. "Anthropological Activities in Connection with the War in England." AJPA, (1918), 1:91-96.

163 KEYS, Donald F. "The American Peace Movement." NHC, (1965), 295-308.

164 KLUCKHOLM, D. "Anthropological Research and World Peace." *Approaches to World Peace, The Fourth Symposium of the Conference on Science, Philosophy, and Religion.* New York, (1944), 143-52.

165 KNOTT, A.J. "East Africa and the Returning Askari: Effect of their War Service." *Quarterly Review,* (1947), 285:98-111.

166 LESSER, Alexander. "War and the State." WAACA, (1968), 92-96.

167 LOPEZ CALERA, N.M. "Antropologia, derecho natural y guerra nuclear." *Revista Estudios Politicos,* (1967), 156:13-27.

168 MacDONALD, Arthur. *War and Criminal Anthropology.* Washington, D.C.: Government Printing Office, 1917.

169 MASSERMAN, Jules, ed. *Science and Psychoanalysis.* v. 6. *Violence and War: With Clinical Studies.* New York: Grune and Stratton, 1963.

170 ——. *Science and Psychoanalysis.* v. 7. *Violence and War.* New York: Grune and Stratton, 1963.

171 McGINN, N.F., E. Harburg, and G.P. Ginsburg. "Response to Interpersonal Conflict by Middle Class Males in Guadalajara and Michigan." AA, (1965), 67:1483-94.

172 MEAD, M. "Anthropological Techniques in War Psychology." *Bulletin of the Menninger Clinic,* (1943), 7:137-40.

173 ——. "War Need Not Mar Our Children." *New York Times Magazine,* (Feb. 15, 1942), 13, 14.

174 MONTAGU, M.F.A. "Racism, the Bomb, and the World's People." *Asia and the Americas,* (1946), 46:533-35.

175 MOSKAS, Charles C. "Civilized Warfare: Why Men Fight." In James P. Spradley, and D. McCurdy, eds., *Conformity and Conflict.* Boston: Little, Brown, 1971.

176 PARDO ZELA, P.G. de. "The Effect of the War on the Indigenous Populations of Peru and Bolivia." AI, (1944), IV: 211-22.

177 PLATT, W. "Natives. The East African Soldier has Proved his Worth in Battle." *National Review,* (1946), 126:41-49.

178 SOLOMON, George F. "Case Studies in Violence." VSE, (1970).

179 TAX, Sol. "War and the Draft." WAACA, (1968), 195-207.

180 ——. "War and the Draft." NH, (1967), 76:10, 54-58.

181 ——, ed. *The Draft: A Handbook of Facts and Alternatives.* Chicago: University of Chicago Press, 1967.

182 UNSEEM, John. "Governing the Occupied Areas of the South Pacific: Wartime Lessons and Peacetime Proposals." *Applied Anthropology,* (1940), 4:1-10.

183 WHITE, Cris W. "Why War? An Attitude Survey: Ideal, Real, Feel." SWAA-AES, (1971).

184 ——. "Why Warfare? An Undergraduate Anthropology Seminar." SWAA-AES, (1971).

185 WILCOX, Walter. "The Loneliness of the Long Distance Soldier." WHR, (1971), 83-89.

C / DEMOGRAPHIC FACTORS

186 COOK, S.F. "Human Sacrifice and Warfare as Factors in the Demography of Pre-colonial Mexico." *Human Biology,* (1946), 18:81-100.

187 DIVALE, William T. "An Explanation for Primitive Warfare: Population Control and the Significance of Primitive Sex Ratios." *The New Scholar,* (1970), 2:173-192.

188 ——. *A Theory of Population Control in Primitive Culture.* M.A. thesis, California State College at Los Angeles, 1971.

189 ——. "An Explanation for Tribal Warfare." Paper presented at the 69th annual meeting of the American Anthropological Association in the session on Warfare, Violence and Law, San Diego, California, 1970.

190 DUMAS, Samuel, and K.O. Vedel-Peterson. *Losses of Life Caused by War.* Oxford: Clarendon Press, 1923.

191 GAILLARD, G. "Les Consequences de la guerre du point de vue demographique." *Societe d'Anthropologie de Paris, Bulletin et Mem,* (1917), 7:197-228; 8:135-156.

192 HANKINS, F.H. "Pressures of Population as a Cause of War." *The Annals of the American Academy of Political and Social Science,* (1940), 198:101-108.

193 HARRIS, Marvin. "Ecology, Demography, and War." In his *Culture, Man, and Nature.* New York: Thomas Crowell, 1971, pp. 200-234.

194 HONE, Fred J. "Military Life: A Neglected Factor in Migration." Paper presented at the 49th annual meeting of the Central States Anthropological Society, Milwaukee, Wisconsin, 1969.

195 HULSE, F.S. "Warfare, Demography, and Genetics." *Eugenics Quarterly*, (1961), 8:185-97.

196 KRZYWICKI, Ludwik. *Primitive Society and Its Vital Statistics*. London: Macmillan, 1934.

197 KULISCHER, Eugene M. *Europe on the Move. War and Population Changes, 1917-47*. New York: Columbia University Press, 1948.

198 PEARL, R. "The Effects of the War on Chief Factors of Population Change." *Science*, (1920), 51:553-56.

199 ——. "A Further Note on War and Population." *Science*, (1921), 53:120-121.

200 PITT-RIVERS, George H.L. *The Clash of Culture and the Contact of Races*. London: George Routledge, 1927.

201 RICHARDSON, Lewis F. *Statistics of Deadly Quarrels*. Pittsburgh: Boxwood Press, 1960.

202 RIVERS, W.H.R. *Essays on the Depopulation of Melanesia*. Cambridge: Cambridge University Press, 1922.

D / BIOLOGICAL ASPECTS

203 "The Aftermath of War." *Eugenical News*, (1916), 1:15.

204 ALLAND, Alexander Jr. "War and Disease: An Anthropological Perspective." WAACA, (1968), 65-75.

205 ——. "War and Disease: An Anthropological Perspective." NH, (1967), 76:10, 58-61, 70.

206 APERT, E. "War and the French Race." [from *Le Monde Medical*, (Jan., 1917)]. *Eugenical News*, (1917), 2:65.

207 BIGELOW, Robert. *The Dawn Warriors: Man's Evolution toward Peace*, London: Hutchinson, 1969.

208 DAVENPORT, C.B., and A.G. Lowe. "Defects found in Drafted Men." *Scientific Monthly*, (1920), 10:5-25, 125-41.

209 FASTEN, Nathen. "War and Eugenics." *Eugenics*, (1929), 2:11-13.

210 GINI, Corrado. "Gli effetti eugenici o disgenici della guerra." *Genus*, (1934), 1:29-42.

211 Harvard University, Department of Anthropology. "Body Build in Relation to Military Function in a Sample of the United States Army." Cambridge, Massachusetts, 1948 (Mimeographed.)

212 HRDLICKA, Ales. "The Effects of War on the American People." *Scientific Monthly*, (1919), 8:542-45.

213 ——. "The Human Losses of the War." AJPA, (1919), 2:77-78.

214 ——. "The Effects of the War on the Race." *Art and Archaeology*, (1918), 7:404-7.

215 HUNT, H.R. *Some Biological Aspects of War.* Eugenics Research Association, Monograph Series, No. 2. New York: Galton, 1930.

216 ——. "Biological Selection in War." *Eugenics*, (1929), 2:3-10.

217 JORDAN, David Starr. "War Selection in the Ancient World." *Scientific Monthly*, (1915), 1:36-43.

218 ——. *The Human Harvest; A Study of the Decay of Races through the Survival of the Unfit.* Boston, American Unitarian Association, 1907.

219 LIVINGSTONE, Frank B. "The Effects of Warfare on the Human Species." WAACA, (1968), 3-15.

220 ——. "The Effects of Warfare on the Biology of the Human Species." NH, (1967), 76:10, 61-65, 70.

221 McINTIRE, R. "The Effect of the War upon the American Child." AJPA, (1919), 2:25-33.

222 MEAD, M. "Warfare is Only an Invention—Not a Biological Necessity." *Asia Magazine*, (1940), 40:402-405.

223 PAUL, Benjamin D. "The Direct and Indirect Biological Costs of War." WAACA, (1968), 76-80.

224 PEARL, Raymond. "Some Biological Considerations about War." AJS, (1941), 46:487-503.

225 SCHNEIDER, David M. "The Social Dynamics of Physical Disability in Army Basic Training." *Psychiatry*, (1947), 10:323-33.

226 SCHURMEIER, H.L. "Congenital Deformities in Drafted Men." AJPA, (1922), 5:51-60.

227 THIEME, Frederick P. "The Biological Consequences of War." WAACA, (1968), 16-21.

228 VERAVAECK, L. "La Revision des tableaux des infirmities et des maladies qui exemptent du service militaire au point du vue anthropologique." *Societe d'Anthropologie de Bruxelles, Bulletin et Memoires*, (1913), 33:205-219.

229 WOODS, Frederick A. "The Biology of War." *Forum*, (1925), 74:533-42.

E / TERRITORIAL & ECONOMIC FACTORS

230 ARAGON, J.O. "Expansion Territorial del imperio Mexicano." *Museo Nacional de Argueologia, Historia, y Etnografia, Anales*, (1931), 7:5-64.

231 ARDREY, Robert. *The Territorial Imperative: A Personal Inquiry into the Animal Origins of Property and Nations.* New York: Atheneum, 1966.

232 BIRMINGHAM, D. *Trade and Conflict in Angola: The Mbundu and Their Neighbours under the Influence of the Portuguese, 1483-1790.* Oxford: Clarendon Press, 1966.

233 CODERE, H. *Fighting with Property: A Study of Kwakiutl Potlatching and Warfare, 1792-1930.* American Ethnological Society Monograph, No. 18, University of Washington Press, 1950.

234 GULLIVER, P.H. "Land Shortage, Social Change, and Social Conflict in East Africa." JCR, (1961), 5:16-26.

235 HENRY, Jules. "National Character and War." AA, (1951), 53:134-135.

236 HOLSTI, R. *Sociological Theory of Sovereignty.* Proceedings of the Institute of International Relations. v. 6. Berkeley, California, 1930.

237 LEVI-STRAUSS, C. "Guerra e commercio entre os indios da America do Sul." *Sao Paulo, Arquivo Municipal, Revista*, (1942), 8:131-146.

238 MEAD, Margaret. *Cooperation and Competition Among Primitive Peoples.* New York: McGraw-Hill, 1937.

239 NAROLL, Raoul. "Imperial Cycles and World Order." *Peace Research Society Papers*, Chicago, (1967), 7:83-101.

240 OTTERBEIN, Keith. "Military Sophistication and Territorial Expansion: A Cross-Cultural Study." Paper presented at the 67th annual meeting of the American Anthropological Association, Seattle, Washington, 1968.

241 ROBINSON, E.V. "War and Economics in History and in Theory." *Political Science Quarterly*, (1900), 15:581-622.

242 SAHLINS, M.D. "Segmentary Lineage: An Organization of Predatory Expansion." AA, (1961), 63:22-45.

243 THOMAS, E.M. *Warrior Herdsmen.* London: Secker and Warburg, 1966.

244 TRIGGER, Bruce G. "War, Trade, and Farming in Lower Nubia." Paper presented at the 62nd annual meeting of the American Anthropological Association, San Francisco, California, 1963.

245 VAYDA, A.P. "Expansion and Warfare Among Swidden Agriculturalists." AA, (1961), 63:346-358.

246 ———. "Revenge and Territorial Conquest in Primitive Warfare." Paper presented at the 69th annual meeting of the American Anthropological Association, San Diego, California, 1970.

247 VISHER, S.S. "Territorial Expansion." *Scientific Monthly,* (1935), 40:440-449.

F / PSYCHOLOGICAL FACTORS

248 ANGELL, Robert S. "The Sociology of Human Conflict." In E. McNeil, ed., *The Nature of Human Conflict.* Englewood Cliffs, New Jersey: Prentice-Hall, 1965, pp. 91-115.

249 ARDREY, Robert. *African Genesis: A Personal Investigation into the Animal Origins and Nature of Man.* New York: Atheneum, 1961.

250 ARONSON, Daniel R. "Social Networks and Social Conflict among the Yoruba." Paper presented at the 49th annual meeting of the Central States Anthropological Society, Milwaukee, Wisconsin, 1969.

251 BARRY, D. "Evolution and the Adaptive Value of War." Unpublished Humanities Honors Program Essay, Stanford University, 1969.

252 BARTELL, G.D. "Review of A.R. Beals' *Divisiveness and Social Conflict.*" *Man,* (1967), 2:639.

253 BEALS, A.R., and B.J. Siegel. *Divisiveness and Social Conflict: An Anthropological Approach.* Stanford: Stanford University Press, 1966.

254 BERKOWITZ, L. *Aggression: A Social Psychological Analysis.* New York: McGraw-Hill, 1962.

255 BERNDT, Ronald M. *Excess and Restraint: Social Control among a (New Guinea) Mountain People.* Chicago: University of Chicago Press, 1962.

256 BROCH, T. "Belligerence among the Primitives." *Japanese Psychological Research,* (1966), 1:33.

257 CLARKSON, J.C., and T.C. Cochran, eds. *War as a Social Institution.* New York: Columbia University Press, 1941.

258 COBLENTZ, S.A. *From Arrow to Atom Bomb: The Psychological History of War.* New York: Beechhurst Press, 1953.

259 COSER, L.A. *Continuities in the Study of Social Conflict.* New York: Free Press, 1967.

260 ———. "Violence and the Social Structure." SPVW, (1963).

261 ———. *The Functions of Social Conflict.* New York: Free Press, 1956.

262 DARLINGTON, H.S. "The Meaning of Head-hunting. An Analysis of a Savage Practice and its Relationship to Paranoia." *Psychoanalytic Review*, (1939), 26: 55-68.

263 FARIS, Robert. *Social Disorganization.* 2nd ed. New York: Ronald Press, 1955.

264 GORER, Geoffrey. "Man Has No Killer Instinct." *New York Times Magazine*, (Nov. 27, 1966).

265 HALL, E.R. "Zoological Subspecies of Man at the Peace Table." *Journal of Mammalogy*, (1946), 27:227-234.

266 HASTINGS, Glover Street. "Man's Inhumanity to Man." *Illinois State Archaeological Society Journal*, (1946), 3:26-27.

267 HRDLICKA, Ales. "War and Race." AA, (1918), 20:239-40.

268 HUXLEY, Julian. "Is War Instinctive—and Inevitable?" *New York Times Magazine*, (Feb. 10, 1946), 59-60.

269 ILFELD, Frederick W., and R.J. Metzner. "Alternatives to Violence: Strategies for Coping with Social Conflict." VSE, (1970), 129-164.

270 ———. "Environmental Theories of Violence." VSE, (1970), 79-96.

271 KERR, Madeline. *Personality and Conflict in Jamaica.* Liverpool: Liverpool University Press, 1952.

272 MEAD, M. "The Psychology of Warless Man." In A. Larson, ed., *Warless World.* New York: McGraw-Hill, 1963, pp. 131-142.

273 ———. "Violence in the Perspective of Culture History." In J. Masserman, ed., *Science and Psychoanalysis.* v. 7. New York: Grune and Stratton, (1963), pp. 92-106.

274 MURPHY, Robert F. "Intergroup Hostility and Social Cohesion." AA, (1957), 59:1018-35.

275 NAROLL, Raoul. "A Tentative Index of Culture-stress." *International Journal of Social Psychiatry*, (1959), 5:107-116.

276 PATRICK, G.T.W. "The Psychology of War." *Popular Science Monthly*, (1915), 87:166-68.

277 PEAR, T.H., ed. *Psychological Factors of Peace and War*. London: Hutchinson, 1950.

278 SPEIER, Hans. *Social Order and the Risks of War*. New York: Stewart, 1952.

279 ———. "The Social Types of War." AJS, (1941), 46:445-54.

280 STAGNER, Ross. "The Psychology of Human Conflict." NHC, (1965), 45-63.

281 THOMAS, W.T. *Primitive Behavior*. New York: McGraw-Hill, 1937.

282 WALLACE, Anthony F. "Psychological Preparations for War." WAACA, (1968), 173-82.

283 ———. "Psychological Preparations for War." NH, (1967), 76:10, 50-54, 70.

284 WALSH, M.N. "Psychic Factors in the Causation of Recurrent Mass Homicide." WHR, (1971), 70-82.

285 WHITNEY, Stephen, and Daniel Katz. "The Social Psychology of Human Conflict." NHC, (1965), 64-90.

286 WHITTEN, N.E. "Review of A.R. Beals' *Divisiveness and Social Conflict*." ASR, (1967), 32:844-45.

287 WOLFGANG, M.E., and F. Ferracuti. *The Subculture of Violence*. London: Social Science Press, 1967.

288 WORSLEY, Peter. "Review of E. Walter *Terror and Resistance*." AA, (1970), 72:1195-97.

F.1 / Aggression

289 BOELKiNS, R. Charles, and J.F. Heiser. "Biological Bases of Aggression." VSE, (1970).

290 CARTHY, J.C., and R.J. Ebling, eds. *The Natural History of Aggression*. New York: Academic Press, 1964.

291 DRIVER, Peter M. "Toward an Ethnology of Human Conflict: A Review of Konrad Lorenz, *On Aggression;* Robert Ardrey, *African Genesis* and *The Territorial Imperative;* and Claire and W.M.S. Russell, *Human Behavior—A New Approach.*" JCR, (1967), 11:361-74.

292 DURBIN, E.F.M., and J. Bowlby. *Personal Aggressiveness of War.* (See pp. 51-150 for a discussion of psychological and anthropological evidence.) New York: Columbia University Press, 1939.

293 DUVALL, Sylvanus M. *War and Human Nature.* New York: Public Affairs Committee, 1947.

294 FREEMAN, Derek. "Human Aggression in Anthropological Perspective." NHA, (1964), 109-19.

295 HAMER, J.H. "An Analysis of Aggression in Two Societies." *Anthropology Tomorrow,* (1956), V: 87-94.

296 HOLLOWAY, Ralph L. "Human Aggression: The Need for a Species-Specific Framework." WAACA, (1968), 29-48.

297 ——. "Human Aggression: The Need for a Species-Specific Framework." NH, (1967), 76:10, 40-44, 69.

298 ——. "Review of Carthy, *Natural History of Aggression.*" AA, (1966), 68:830-32.

299 KAHN, M.W., and W.E. Kirk. "The Concept of Aggression: A Review and Reformulation." *Psychological Record,* (1968), 18:559.

300 LORENZ, Konrad. *On Aggression.* New York: Bantam, 1967.

301 McKELLAR, P. "The Emotion of Anger in the Expression of Human Aggressiveness." *British Journal of Psychology,* (1949), 39:148.

302 McNEIL, Elton B. "The Nature of Aggression." NHC, (1965), 145-54.

303 ——. "Psychology and Aggression." JCR, (1959), 3:195-293.

304 MELGES, Frederick T., and R.F. Harris. "Anger and Attack: A Cybernetic Model of Violence." VSE, (1970), 97-128.

305 MONTAGU, M.F.A., ed. *Man and Aggression.* New York: Oxford University Press, 1968.

306 NICOLAI, Georg Friedrich. *The Biology of War.* Translated by Constance A. Grande and Julian Grande. New York: Century Company, 1918.

307 PALMER, Stuart. "Murder and Suicide in Forty Nonliterate Societies." *Journal of Criminal Law, Criminology, and Police Science,* (1965), 56:320-24.

308 SCOTT, J.P. *Aggression.* Chicago: University of Chicago Press, 1958.

309 SOLOMON, George F. "Psychodynamic Aspects of Aggression, Hostility, and Violence." VSE, (1970).

310 STRATTON, George Malcolm. *The Control of the Fighting Instinct.* International Conciliation, No. 73, 1913.

311 WHITE, William A. "War and Human Nature." In R.E. Park, ed., *Introduction to the Science of Sociology.* Chicago: University of Chicago Press, 1921, pp. 594-98.

F.2 / Animal War

312 CARPENTER, C.R. "The Contribution of Primate Studies to the Understanding of War." WAACA, (1968), 49-58.

313 EIBL-EIBESFELDT, Ivenaus. "The Fighting Behavior of Animals." *Scientific American,* (1961), 205: 112-16, 119-20, 122.

314 FRIEDMANN, H. "Animal Aggression and its Implications for Human Behavior." WHR, (1971), 12-23.

315 JAY, Phyllis. "Dominance." Paper presented at the 62nd annual meeting of the American Anthropological Association, San Francisco, California, 1963.

316 KROTT, Peter. "A Key to Ferocity in Bears." NH, (1961), 71:64-71.

317 LORENZ, Konrad. "Ritualized Fighting." NHA, (1964).

318 SUTTLES, W. "Subhuman and Human Fighting." *Anthropologica,* (1961), 3:148-63.

319 TINBERGEN, N. "On War and Peace in Animals and Man." *Science,* (1968), 160:1411-18.

320 ———. *The Study of Instinct.* Oxford, 1951.

321 ———. "Fighting and Threat in Animals." *New Biology,* (1953), 14:9-24.

322 WRIGHT, Quincy. "Animal Warfare." In Q. Wright, *A Study of War.* Chicago: University of Chicago Press, 1942, pp. 42-48.

323 WYNNE-EDWARDS, V.C. *Animal Dispersion in Relation to Social Behavior.* London: Oliver and Boyd, 1962.

324 ———. "Self Regulation System in Populations of Animals." *Science,* (1962), 147:1543-47.

G / FEUDING

325 BEALS, A. "Conflict and Interlock Festivals in a South Indian Region."
     Journal of Asian Studies, (1961), 231:99-113.

326 ——. "Cleavage and Internal Conflict: An Example from India." JCR,
     (1961), 5:27-34.

327 ——. "Culture Change and Social Conflict in a South Indian Village."
     Unpublished Ph.D. dissertation, University of California, Berkeley,
     1954.

328 BELL, F.L. "Organized Violence in a Primitive Community." *Mankind*,
     (1939), 2:186-87.

329 BENEDICT, B. "Factionalism in Mauritian Villages." BJS, (1957),
     8:328-42.

330 BOISSEVAIN, J. "Factions, Parties, and Politics in a Maltese Village." AA
     (1964), 66:1275-87.

331 COLSON, Elizabeth. "Social Control and Vengeance in Plateau Tonga
     Society." *Africa*, (1953), 23:199-211.

332 DANDLER, Jorge. "Patronage and Violence among Peasants in the
     Cochabamba Valley, Bolivia." Paper presented at the 69th annual
     meeting of the American Anthropological Association, San Diego,
     California, 1970.

333 DOZIER, E.D. "Factionalism at Santa Clara Pueblo." *Ethnology*, (1966),
     5:172-85.

334 ELEZI, I. "Sur la Vendetta en Albanie." *Studia Albanica*, (1966),
     3:305-18.

335 FENTON, W.N. *Factionalism at Taos Pueblo.* Bureau of American
     Ethnology, Bulletin No. 164. Washington, D.C.: Government Printing
     Office, 1957.

336 FIRTH, W.N., et. al. "Factions in Indian and Overseas Indian Societies."
     BJS, (1957), 8:291-342.

337 FRENCH, David. *Factionalism in Isleta Pueblo.* American Ethnological
     Society, Monograph No. 14. Washington: University of Washington
     Press, 1948.

338 GLASSE, R.M. "Revenge and Redress among the Huli: A Preliminary
     Account." *Mankind*, (1959), 5:273-89.

339 HADDON, A.C. "Quarrels and Warfare." RCAETS, (1908), 6:189-91.

340 HASLUCK, Margaret. "The Albanian Blood Feud." LWSAC, (1967), 381-408.

341 JONES, V.C. *The Hatfields and the McCoys.* Chapel Hill, N.C.: University of North Carolina Press, 1948.

342 KROEF, J.M. van der. "Social Conflict and Minority Aspirations in Indonesia." AJS, (1949-50), LV: 450-463.

343 LARSON, G.F. "Warfare and Feuding in the Ilaga Valley." In *Working Papers in Dani Ethnology*, v. 1. Holland: Bureau of Native Affairs, 1962.

344 LASSWELL, Harold D. "Feuds." *Encyclopaedia of the Social Sciences,* (1931), 6.

345 LEWIS, O., and H.S. Dhillan. *Group Dynamics in a North-Indian Village: A Study of Factions.* New Delhi, India: Programme Evaluation Organization Planning Commission, Government of India, n.d.

346 MAYER, A.C. "Factions in Fiji Indian Rural Settlements." BJS, (1957), 8:317-28.

347 MEAD, M. "Cultural Factors in the Cause of Prevention of Pathological Homicide." *Bulletin of the Menninger Clinic*, (1964), 28:11-22.

348 MORRIS, H.S. "Communal Rivalry Among Indians in Uganda." BJS, (1957), 8:306-17.

349 NASH, J.C. "Death as a Way of Life: The Increasing Resort to Homicide in a Maya Indian Community." AA, (1967), 69:5, 455-70.

350 OTTERBEIN, Keith and Charlotte. "An Eye for an Eye, A Tooth for a Tooth: A Cross-cultural Study of Feuding." AA, (1965), 67:1470-82.

351 PELTO, Pertti, and J. MacGregor. "Competition and Hostility in Little Communities." Paper presented at the 62nd annual meeting of the American Anthropological Association, San Francisco, California, 1960.

352 PETERS, E.L. "Some Structural Aspects of the Feud among the Camel-Herding Bedouin of Cyrenaica." *Africa*, (1967), 37:261-82.

353 POCOCK, D. "The Bases of Faction in Gujerat." BJS, (1957), 8:295-306.

354 POSPISIL, Leopold J. "Feud." *International Enclyclopedia of the Social Sciences.* v. 5. 1968.

355 SCHWARTZ, N.B. "Goal Attainment through Factionalism: A Guatemalan Case." AA, (1969), 71:1088-1108.

356 SIEGEL, B.J., and A.R. Beals. "Pervasive Factionalism." AA, (1960), 62:394-417.

357 ——. "Conflict and Factionalist Dispute." JAI, (1960), 90:107-17.

358 STIRLING, A.P. "A Death and a Youth Club: Feuding in a Turkish Village." *Anthropological Quarterly*, (1960), 33:51-75.

359 TOBIN, J.E. *An Investigation of the Socio-Political Schism on Majuro Atol*, 1953 (Mimeographed).

360 TORRES-TRUEBA, H. "Factionalism in a Mexican Municipio." *Sociologus*, (1969), 19:134-52.

361 VAN VELZEN, H.U.E., and W. van Wetering. "Residence, Power Groups and Intra-Societal Aggression: An Inquiry into the Conditions Leading to Peacefulness within Non-Stratified Societies." *International Archives of Ethnography*, (1960), 49:169-200.

362 WESTERMARCK, E. "The Blood-Feud among Some Berbers of Morocco." In E. Westermarck, ed., *Essays to Seligman*. London, 1934.

363 YADARA, J.S. "Factionalism in a Haryana Village." AA, (1968), 70:898-910.

364 ZACHARIAS, R. "Die Blutrache im deutschen Mittelalter." *Zeitschrift fur deutsches Altertum und deutsche Literatur*, (1962), 91:167-201.

H / PEACE & PEACEMAKING

365 "Sioux Treaty of 1868." IH, (1970), 3:13-17.

366 BACDAYAN, Albert S. "The Peace-Pact Celebration: Review and Revitalization of Inter-Village Law among the Kalinga Ingorots of Northern Luzon, The Philippines." Paper presented at the 67th annual meeting of the American Anthropological Association, Seattle, Washington, 1968.

367 CHAGNON, N.A., and Timothy Asch. *The Feast*. Motion picture on Yanomamo warfare alliances. Atomic Energy Commission, 1968.

368 CHAPMAN, Anne M. "Raices y consecuencias de la guerra de los Aztecas contra los tepanecas de azcapotzalco." *Acta anthropologie*, (1959), 2(1):4.

369 COOK, Blanche Viesen. *Bibliography on Peace Research in History*. American Bibliographical Center, Reference Series, No. 11. Santa Barbara, California: Clio Press, 1969.

370 DYK, Walter. "Review of G. MacGregor, *Warriors Without Weapons.*" AA, (1947), 49:279.

371 HOBHOSE, L.T. "The Simplest Peoples. Part II: Peace and Order among the Simplest Peoples." BJS, (1956), 7:96-119.

372 JANIS, Irving L., and Daniel Katz. "The Reduction of Intergroup Hostility." JCR, (1959), 3:85-100.

373 JORDAN, Virgil. "The Peace Myth." *American Mercury*, (1924), 5:1-9.

374 KAPPLER, C.J. *Indian Affairs: Laws and Treaties.* 4 Vols., Washington: Government Printing Office, 1903-29.

375 KELLER, A.G. *Through War to Peace.* New York, 1918.

376 LA FLESCHE, F. "War Ceremony and Peace Ceremony of the Osage Indians." BBAE, (1939), CI: 1-280.

377 LARSON, Arthur. *A Warless World.* New York: McGraw-Hill, 1963.

378 LEVINE, Robert A. "An Anthropological Study of War and Peace." Paper presented at the 69th annual meeting of the American Anthropological Association, San Diego, California, 1970.

379 LOWIE, R. "Compromise in Primitive Society." *International Social Science Journal,* (1963), 2:182.

380 MacGREGOR, Gordon. *Warriors Without Weapons.* Chicago: University of Chicago Press, 1946.

381 MALINOWSKI, B. "The Deadly Issue." *Atlantic Monthly,* (1936), 158:559-69.

382 MATHUR, M.E.F. "The Jay Treaty and Confrontation at St. Regis Boundary." IH, (1970), 3:37-40.

383 MEAD, M. "Alternatives to War." WAACA, (1968), 215-28.

384 ———. "Alternatives to War." NH, (1967), 76:10, 65-69, 70.

385 ———, and Rhoda Metraux. "The Participation of Anthropologists in Research Relevant to Peace." Manuscript prepared for the Center for Research in Conflict Resolution and the American Anthropological Association, 1962.

386 MURPHY, Gardner. *Human Nature and Enduring Peace.* New York: Houghton Mifflin, 1945.

387 NAROLL, Raoul. "Deterrence in History." In D. Pruitt and R. Snyder, eds., *Theory and Research on the Causes of War.* Englewood Cliffs: Prentice-Hall, 1969.

388 ———. "Does Military Deterrence Deter?" *Trans-Action*, (1966), 3:14-20.

389 ———. *Warfare, Peaceful Intercourse, and Territorial Change: A Cross-cultural Survey.* Northwestern University and Institute for Cross-Cultural Studies, n.d. (Mimeograph.)

390 ———, Vern Bullough, and Frada Naroll. *Military Deterrence in History.* Albany, New York: State University of New York Press, 1973.

391 NEWCOMBE, Hanna. *Bibliography on War and Peace.* Dundas, Ontario: Canadian Peace Research Institute, 1963.

392 NUMELIN, Ragnar. *The Beginnings of Diplomacy.* London: Oxford University Press, 1950.

393 ———. "Messengers, Heralds, and Envoys in Savage Societies." *Transactions of the Westermarch Society.* v. 1. Gothenburg, Sweden, 1947.

394 OTTLEY, R.L. "Peace." *Hastings Encylopedia of Religion and Ethics,* (1917), 10.

395 PARKER, A.C. "The Peace Policy of the Iroquois." SW, (1911), XL: 691-99.

396 PEAR, T.H. "Peace, War, and Cultural Patterns." *John Rylands Library Bulletin,* (1948), 31:120-47.

397 PERRY, W.J. "The Peaceable Habits of Primitive Communities . . . An Anthropological Study of the Golden Age." *Hibbert Journal,* (1917), 16:28-46.

398 SCHEFFLER, H.W. "The Social Consequences of Peace on Choiseul Island." *Ethnology,* (1964), 3:398-403.

399 TEFFT, Stanton K. "Peacemaking Procedures of Stateless Societies and Modern States: A Comparison." SWAA-AES, (1971).

400 ———. "Warfare Resolutions among Primitive Peoples: Some Preliminary Observations." Paper presented at the 68th annual meeting of the American Anthropological Association, New Orleans, Louisiana, 1969.

401 THOMSON, D., E. Meyer, and A. Briggs. *Patterns of Peacemaking.* London, 1945.

402 THOMSON, Warren S. *Population and Peace in the Pacific.* Chicago: University of Chicago Press, 1946.

403 VAN VELZEN, H.U.E. Thoden, and W. van Wetering. "Residence, Power Groups, and Intra-social Aggression: An Inquiry into the Conditions Leading to Peacefulness within Non-Stratified Societies." *International Archives of Ethnography,* (1960), 49:169-200.

404 WALLACE, Paul A. *The White Roots of Peace.* Philadelphia: University of Pennsylvania Press, 1946.

I / COLONIAL PACIFICATION

405 BACDAYAN, Albert S. "Unite and Rule: The Control of Headhunting in the Northern Luzon Mountains of the Philippines by the American Colonial Authorities." Paper presented at the 68th annual meeting of the American Anthropological Association, New Orleans, Louisiana, 1969.

406 BADEN, Powell. *The Matabele Campaign, 1896. Narrative of the Campaign in Suppressing the Native Rising in Matabele and Mashonaland.* 1897.

407 BEACH, W.W. "Chastisement of the Yamasees: An Incident of the Early Indian Wars in Georgia." In W. Beach, *Indian Miscellany.* Albany: Munsell, 1877.

408 BLANKENSTEIN, M. van. "War, the Stabilizer of the Dutch Empire." AR, (1940), XXXVI: 793-798.

409 BOURKE, J.G. "Mackenzie's Last Fight with the Cheyennes." *Journal of the Military Service Institution of the United States.* (1890), XI:29-49, 198-221.

410 BOYLE, Lieut. W.H. *Personal Observations on the Conduct of the Modoc War.* Pacific MSS, no. A96. Bancroft Library, University of California, Berkeley.

411 BURNS, R.I. "A Jesuit in the War against the Northern Indians." RACHS, (1950), LXI, 9-54.

412 CALLWELL, C.E. *Small Wars: Their Principles and Practice.* 3rd ed. London, 1906.

413 COLBY, Elbridge. "How to Fight Savage Tribes." *American Journal of International Law,* (April 1922), 280-84.

414 DOWNEY, Fairfax. *Indian Fighting Army.* New York: Charles Scribner's Sons, 1941.

415 FARRER, J.A. "Savage and Civilized Warfare." JAI, (1880), 9:358-68.

416 FINOT, E. *Historia de la conquista del Orienti Boliviano.* Buenos Aires, 1939.

417 FISHER, Don C., and Ben Wright. *A Brief History of the Modoc Indians and Their War Fought in 1872-73.* Lava Beds National Monument, Tulelake, California.

418 GRAHAM, R.B.C. *The Conquest of New Granada.* London: W. Heinemann, (1922).

419 HELPS, Arthur. *Spanish Conquest in America.* 4 vols. London: J.W. Parker & Sons, 1855.

420 HENEKER, Lieutenant-Colonel W.C.G. *Bush Warfare.* London, 1907.

421 HUNTER, Monica. *Reaction to Conquest: Effects of Contact with Europeans on the Pondo of South Africa.* London: Oxford University Press, 1936.

422 JONES, O. *Pueblo Warriors and Spanish Conquest.* Norman: University of Oklahoma Press, 1966.

423 KEESING, F.M. and Maria. *Taming Philippine Headhunters: A Study of Government and of Cultural Change in Northern Luzon.* London: George Allen and Unwin, 1934.

424 KENNEDY, G.A. "The Last Battle." *Alberta Folklore Quarterly,* (1945), 1:57-60.

425 MANRING, B.F. *The Conquest of the Coeur d'Alenes, Spokanes and Palouses.* Spokane: Inland Printing Co., 1912.

426 NEQUATEWA, E. "A Mexican Raid on the Hopi Pueblo of Oraibi." P, (1944), XVI:44-52.

427 NICOLAI, H. "Conflits entre groupes africains et décolonisation au Kasai." *Revue de L'Université de Bruxelles,* (1959), 12:131-144.

428 POWELL, Philip Wayne. *Soldiers, Indians and Silver. The Northward Advance of New Spain, 1550-1600.* Berkeley & Los Angeles: University of California Press, 1952.

429 ———. "Spanish Warfare against the Chichimecas in the 1570's." HAHR, (1944), 24:580-604.

430 REEVE, F.D. "Navajo Spanish Wars: 1680-1720." NMHR, (1958), XXXIII:205-31.

431 ROHDEN, L. von. *History of the Rhenish Missionary Association.* (Report on domicilation of natives in South Africa by the missionaries.) Barmen, 1888.

432 ROSS, John E. *Narrative of an Indian Fighter.* Pacific MSS, Bancroft Library, University of California, Berkeley, n.d.

433 ROWLEY, C. "Culture Clash — Lime and Gunpowder." *South Pacific,* (1952), 6:513-15.

434 SAMAYOA, Chinchilla C. "Causas que mas influyeron en las derrotas de los ejercitos indigenas durante las querras de la conquista." Mexico, *Cuadernos Americanos,* (1960), 19:133-49.

435 ———. "Causas de las derrotas indigenas durante la conquista." Servilla, *Estudios Americanos,* (1959), 18:245-60.

436 SAVARY, H. "Conquete des anciens Chiliens par les Peruviens, au temps des Incas." ICA, (1878), II,i:361-63.

437 SPRAGUE, J.T. *The Origin, Progress and Conclusion of the Florida War.* N.T., 1848.

438 THOMAS, A.B. *Forgotten Frontiers; a Study of the Spanish Indian Policy of Don Juan Bautista de Anza, Governor of New Mexico, 1777-1787.* Norman: University of Oklahoma Press, 1932.

439 THURNWALD, R. "The Price of the White Man's Peace." *Pacific Affairs,* (1936), 9:358.

440 TODD, J.A. "Native Offences and European Law in South-West New Britain." *Oceania,* (1935), 5:437-60.

441 TREMEARNE, Major A.J.N. *The Tailed Head Hunters of Nigeria.* Philadelphia: Lippincott, 1912.

442 TREUBA y Cosio, T. de. *History of the Conquest of Peru by the Spaniards.* Edinburgh, 1830.

443 WILLIS, William Shedrick. *Colonial Conflict and the Cherokee Indians, 1710-1760.* Ph.D. thesis, Columbia University, 1955.

J / LAW & ORDER

444 *Sarawak, The Criminal Procedure Code, (1932).* Kuching, 1932.

445 BOHANNAN, P.J. *African Homicide and Suicide.* Princeton, New Jersey: Princeton University Press, 1960.

446 BROWN, Donald. "Law and Disorder in an Indian Pueblo." Paper presented at the 69th annual meeting of the American Anthropological Association, San Diego, 1970.

447 CHASTEEN, E. "Public Accommodations: Social Movements in Conflict." *Phylon,* (1969), 30:234-50.

448 COHN, B.S. "Anthropological Notes on Disputes and Law in India." AA, (1965), 67 (6, pt. 2):82-122.

449 DEMETRACUPOULOU, D. "Wintu War Dance." PPSC, (1939), VI, iv:141-143.

450 DEVEREUX, G., and E.M. Loeb. "Some Notes on Apache Criminality." *Journal of Criminal Psychopathology*, (1943), IV:424-30.

451 GRINNELL, G.B. *The Punishment of the Stingy.* New York: Harper and Bros., 1901, pp. 127-235.

452 GULLIVER, P. *Social Control in an African Society.* Boston: Boston University Press, 1963.

453 HARGRETT, Lester. *A Bibliography of the Constitutions and Laws of the American Indians.* Cambridge: Harvard University Press, 1947, pp. 106-9.

454 HEIZER, R.F. "Executions by Stoning among the Sierra Miwok and Northern Paiute." PKAS, (1954), XII:45-54.

455 HENDERSON, Dan. "Settlement of Homicide Disputes in Sakya (Tibet)." AA, (1964), 66:1099-1148.

456 HOEBEL, E.A. "Law-ways of the Primitive Eskimos." *Journal of Criminal Law and Criminology.* (1941), XXXI:663-83.

457 ———. "The Political Organization and Law-ways of the Comanche Indians." MAAA, (1940), LIV:1-149.

458 HOGBIN, H. Ian. "Social Reaction Crime: Law and Morals in the Shouten Islands, New Guinea." JAI, (1948), 68:232-62.

459 ———. *Law and Order in Polynesia.* New York: Harcourt Brace, 1934.

460 HOPKINS, Elizabeth. "The Politics of Crime: Patterns of Aggression and Control in a Colonial Context." Paper presented at the 67th annual meeting of the American Anthropological Association, Seattle, 1968.

461 HOWELL, P.P. "Observations on the Shilluk of the Upper Nile. The Laws of Homicide and the Legal Functions of the Reth." *Africa,* (1952), 22:109-19.

462 MEEK, Charles K. *Law and Authority in a Nigerian Tribe.* London: Oxford University Press, 1950.

463 NADER, L. "Choices in Legal Procedure: Shia Moslem and Mexican Zapotec." AA, (1965), 67:394-99.

464 OPLER, Morris E. "Review of G. Roheim, *War, Crime and the Covenant.*" AA, (1946), 48:459-61

465 OTTERBEIN, Keith. "Conflict and Communication: The Social Matrix of Obeah." *Kansas Journal of Sociology*, (1965), 3:112-28.

466 PARK, Robert E. "The Social Function of War." AJS, (1941), 46:551-70.

467 ROHEIM, Geza. *War, Crime and the Covenant*. Monticello, New York: Medical Journal Press, 1945.

468 SCHNEIDER, Joseph. "Is War a Social Problem?" JCR, (1959), 3:353-60.

469 Winnebago Tribe of Indians. *Laws and Regulations*. Omaha: Tribal Council, 1868.

K / HEAD HUNTING

470 "Head-Hunting, a Rite Now Given Over, by Which the Once Wild Tribes of the Philippines Canceled the 'Debt of Life.'" *Asia*, (1924), 24:629-96.

471 AUSTEN, L. "Karigara Customs." *Man*, (1923), 23:35-36.

472 BARTON, R.F. "Hunting Soul-stuff; the Motive behind Head-taking as Practised by Ifugaos of the Philippines." *Asia*, (1930), 30:188-95, 225-26.

473 BOCK, C. *The Head-hunters of Borneo*. London, 1881.

474 COTLOW, Lewis. *Amazon Head-hunters*. London: Holt, 1953.

475 CUMMINGS, Lewis V. *I Was a Head-hunter*. Boston: Houghton Mifflin, 1941.

476 DARLINGTON, H. "The Meaning of Head Hunting; an Analysis of a Savage Practice and Its Relationship to Paranoia." *Psychoanalytic Review*, (1939), 26:55-68.

477 DOWNS, R.E. "Head Hunting in Indonesia." BTLV, (1955), 3:40-70.

478 DURHAM, M.E. "Head Hunting in the Balkans." *Man*, (1923), 23:19-21.

479 FISCHER, H.T. "On Head Hunting." BTLV, (1955), 3:274-80.

480 FURER-HAIMENDORF, C. "The Last Head-hunting Feast of the Konyak Nagas of Assam." *The Illustrated London News*, (February 5, 1938), 208-11, 234.

481 GIGLIOLY, E.H. "I cacciatori di teste alla Nuova Guinea." *Archivio per L'Anthropologia e L'Ethnologia*, (1896), 26:311-18.

482 GILZEN, K.K. "Cheloviecheskaia golova, kak voennyi trofei, u indieitsev plemeni munduruke." *Akademiia Nauk SSSR.* (Muzei Antropologii i etnografii), (1917-25), 5:351-58.

483 HADDON, A.C. *Head-Hunters: Black, White, and Brown.* London, 1901.

484 ———. "The Ingeri Head Hunters of New Guinea." *Internationales Archiv fur Ethnologie,* (1891), 4:177-81.

485 HILL, Jasper (Big White Owl). ". . . And They Called Us Savages." *Native Voice,* (1950), 4:2, 15.

486 HUTTON, J.H. "Head Hunting." *Man in India,* (1930), 10:207-15.

487 KARSTEN, R. *Headhunters of Western Amazonas: The Life and Culture of the Jivaro Indians of Eastern Ecuador and Peru.* Helsinki, Societas Scientiarum Pennica, Commentationes Humanarum Litterarum, (1935), 8 (1).

488 KROEF, Justus M. van der. "Headhunting Customs of Southern New Guinea." *United Asia,* (1955), 7:159-63.

489 ———. "Some Head-hunting Traditions of Southern New Guinea." AA, (1952), 54:221-35.

490 LING, Shun-Sheng. "The Head-hunting Ceremony of the Wa Tribe and That of the Formosan Aborigines." Taipei, Formosa University, Department of Anthropology and Archaeology *Bulletin,* (1953), 2:1-9.

491 McCARTHY, F.D. "Head-hunters of Oceania." *Australian Museum Magazine,* (1959), 13:76-80.

492 MIDDLEKOOP, Pieter. *Head Hunting in Timor and Its Historical Implications.* Oceania Linguistic Monographs No. 8. Sydney: University of Sydney, 1963.

493 MYTINGER, C. *New Guinea Headhunt.* New York: MacMillan, 1946.

494 NYANDOH, R. "Head-hunting Revenge." (Land Dayak.) *Sarawak Museum Journal,* (1958), 8:732-35.

495 OKADA, Yuzuru. "Head-hunting in the Atayal Tribe of Formosa." (text in Japanese.) *Minzokugaku-Kenkyu,* (1935), 1:123-27.

496 PILARES, Polo E. "Una 'practica' de caracter etnico atribuida al incanato." *Cuzco Instituto Arqueologico, Revista,* (1937), 3:72-74.

497 REED, S.W. "Headhunting in New Guinea." AA, (1952), 54:576-77.

498 RILEY, E. Baxter. *Among Papuan Headhunters.* Philadelphia: Lippincott, 1925.

499 ———. "Dorro Head-hunters." *Man,* (1923), 23:33-35.

500 SAULNIER, Tony. *Headhunters of Papua.* New York: Crown, 1963.

501 SHARP, H.H. "Gumakari People of the Suki Creek, New Guinea." *Man,* (1934), 34:97-98.

502 STIRLING, M.W. "Head Hunters of the Amazon." *Scientific Monthly,* (1933), 36:264-66.

503 THOMSON, Donald. "With the Headhunters of Dutch New Guinea." *Exploration,* (1961), 1:58-65.

504 TREMEARNE, Major A.J.N. "The Kagoro and Other Nigerian Head-Hunters." JAI, (1923), 42:136-99.

505 UP de GRAFF, F.W. *Head Hunters of the Amazon: Seven Years of Exploration and Adventure.* Garden City, New York: Doubleday, 1923.

506 Van Der KROEF, J.M. "Head Hunting Customs of Southern New Guinea." *United Asia,* (1955), 7:159-63.

507 WILLCOX, Cornelis De Witt. *The Head Hunters of Northern Luzon.* Kansas City: Franklin Hudson Co., 1912.

508 WIRZ, Paul. "Headhunting Expeditions of the Tugeri into the Western Division of British New Guinea." *Tijdschrift van de Instituut voor Taal-, Land- en Volkenkuude,* (1933), 73:105-22.

509 WORCESTER, Dean C. "Head-Hunters of Northern Luzon." *National Geographic,* (1912), 23:833-930.

510 ZEGWAARD, Rev. G.A. "Head-hunting Practices of the Asmat of Netherlands New Guinea." AA, (1959), 61:1020-41.

L / CANNIBALISM

511 BEAVER, W.N. "Some Notes on the Eating of Human Flesh in Western Papua." *Man,* (1914), 14:145-47.

512 BOCCASSINO, R. "La vendetta del sangue praticata dagli Acioli dell'Uganda: Riti e cannibalismo queereschi." *Anthropos,* 57:357-73.

513 BROWN, Jennifer. "The Care and Feeding of Windigos: A Critique." AA, (1971), 73:20-22.

514 COWAN, Jas. "The Last of the Cannibals." *Lone Hand,* (1909), 5:568-73.

515 CRAWFORD, John. "On Cannibalism in Relation to Ethnology." *Ethnological Society of London, Transactions,* (1866), 4:105-24.

516 DESCAMPS, Paul. "Le Cannibalisme, ses causes et ses modalites." *L'Anthropologie,* (1925), 35:321-44.

517 DUPEYRAT, A. "Le Cannibalisme en Papouasie et ailleurs." *Annales de N.-D. du Sacre-Coeur,* (September 1958), 194-97.

518 EVANS-PRITCHARD, E. "Cannibalism: A Zande Text." *Africa,* (1956), 26:73-74.

519 GARN, S.M., and W. Black. "The Limited Nutritional Value of Cannibalism." AA, (1970), 72:106.

520 GIORGETTI, F. "Il cannibalismo dei Niam Niam." *Africa,* (1957), 27:178-86; (1958), 28:167.

521 GLASSE, R.M. "Cannibalism in the Kuru Region of New Guinea." *Transactions of the New York Academy of Science,* Series II, (1967), 29:748-54.

522 GREFFRATH, H. "Der Cannibalismus auf den Fidschi-Inseln." *Ausland,* (1891), 64:339.

523 HASZARD, H.D.M. "Notes on Some Relics of Cannibalism." *Transactions of the New Zealand Institute,* (1890), 22:104-5.

524 HAY, Thomas H. "The Windigo Psychosis: Psychodynamic, Cultural, and Social Factors in Aberrant Behavior." AA, (1971), 73:1-19.

525 HOGG, G. *Cannibalism and Human Sacrifice.* London: Robert Hale, 1958.

526 KARLIN, A.M. "Malaita, die Menschenfresser-Insel." *Der Erdball,* (1930), 4:184-87.

527 ——. "Salomoniens Cannibales." *Revue d'Ethnologie,* (1887), 6:76.

528 KOCH, Klaus-Friedrich. "Cannabalistic Revenge in Jale Warfare." In James P. Spradley and D. McCurdy, eds., *Conformity and Conflict.* Boston: Little, Brown, 1971.

529 LOEB, E.M. *The Blood Sacrifice Complex.* American Anthropological Association, Memoir No. 30, 1923.

530 METRAUX, A. "A Cross-cultural Survey of South American Indian Tribes. Warfare, Cannibalism, and Human Trophies." HSAI, (1949), 5:383-409.

531 MOONEY, J. "Our Last Cannibal Tribe." *Harper's Magazine.* (September, 1901).

532 PARKER, Seymour. "The Wiitiko Psychosis in the Context of Ojibwa Personality and Culture." AA, (1960), 62:603-23.

533 PRATT, A.E. *Two Years Among New Guinea Cannibals.* London: Seeley and Co., 1906.

534 RANDALL, Mark. "Comment on 'The Limited Nutritional Value of Cannibalism.' " AA, (1971), 73:269.

535 RICE, Arthur P. "Cannibalism in Polynesia." *American Antiquarian,* (1910), 32:77-84.

536 ———. "Polynesian Cannibalism." AA, (1909), 11:487.

537 ROHRL, V.J. "A Nutritional Factor in the Windigo Psychosis." AA, (1970), 72:97-101.

538 SCHMITZ, Carl A. "Kannibalismus und Todeszauber auf Neuguinea." *Umschau in Wissenschaft und Technik,* (1960), 60:590-603, 619-20.

539 ———. "Zum Problem des Kannibalismus im Nordlichen Neuguinea." *Paideuma,* (1958), 6:381-410.

540 SHANKMAN, Paul. "Le Roti et le Bouilli: Levi-Strauss' Theory of Cannibalism." AA, (1969), 71:54-69.

541 STADEN, Hans. "Killing and Eating One's Enemy in Sixteenth Century Brazil." PH, (1953), 494-96.

542 STEEL, Thos. "Cannibals and Cannibalism." *Victorian Naturalist,* (1893), 10:4-10, 26-30.

543 STEINMETZ, Rudolf S. "Endokannibalismus." MAGW, (1896), 26:1-60.

544 VAYDA, Andrew P. "On the Nutritional Value of Cannibalism." AA, (1970), 72:1462-63.

545 ———. "Maori Women and Maori Cannibalism." *Man,* (1960), 60:70-71.

546 VOLHARD, Ewald. *Il Cannibalismo.* Torino: Einaudi, 1949.

547 ———. *Kannibalismus.* Stuttgart, 1939.

548 WALENS, S., and R. Wagner. "Pigs, Proteins, and People-eaters." AA, (1971), 73:269-70.

549 WATT, W. "Cannibalism as Practised on Tanna, New Hebrides." JPS, (1895), 4:226-30.

550 WEEKS, Rev. J.H. *Among Congo Cannibals.* London, 1913.

M / SCALPING & WAR TROPHIES

551 "Scalping." *Masterkey*, (1940), 14:131.

552 "Scalping." *Masterkey*, (1944), 18:130.

553 "Trophy Heads Exhibited." *Field Museum News*, (1940), 9:7.

554 ACKERKNECHT, E.H. "Origin and Distribution of Skull Cults." *Ciba Symposia*, (1944), 5:1654-61.

555 ———. "Head Trophies and Skull Cults in the Old World." *Ciba Symposia*, (1944), 5:1662-69.

556 ———. "Head Trophies in America." *Ciba Symposia*, (1944), 5:1670-75.

557 DIECK, A. "Archaologische Belege für den Brauch des Skalpierens in Europa." (Archaeological Evidence for Scalping in Europe.) *Neue Ausgrabungen und Forschungen in Niedersachsen*. V.H. Hildesheim, Lax, (1969), 359-71.

558 FLETCHER, Alice C. "The Significance of the Scalp-lock." JAI, (1907), 27:436-50.

559 FORRER, R. "Crane-trophee scalpeneolithique trouve a Archenheim, environs de Strasbourg." *Revue Anthropologique*, (1922), 32:244.

560 FRIEDERICI, Georg. *Skalpieren und ähnliche Kriegsgebrauche in Amerika*. Braunschweig: Druck von Friedrich Vieweg, 1906.

561 ———. "Scalping in America." Washington, *Smithsonian Institution, Report for 1906*, (1907), 423-38.

562 ———. "Scalping Among the North American Indians." Washington, *Smithsonian Institution Annual Report for 1906*, (1907), 423-38.

563 GRINNELL, G.B. "Coup and Scalp Among the Plains Indians." AA, (1910), 12:296-310.

564 GUSINDE, M. "Ursprung und Verbreitung des Schadelkultes." *Ciba Zeitschrift*, (1937), 5:1678-82.

565 ———. "Schadelkult in der Alten Welt." *Ciba Zeitschrift*, (1937), 5:1683-89.

566 ———. "Kopftrophaen in Amerika." *Ciba Zeitschrift*, (1937), 5:1690-96.

567 ———. "Skalp und Skalpiern in Nordamerika." *Ciba Zeitschrift*, (1937), 5:1700-5.

568 HADDON, A.C. "Stuffed Human Heads from New Guinea." *Man*, (1923), 23:36-39.

569 HAGEN, V.W. "Shrunken Heads." NH, (1952), 61:128-30, 141.

570 HERMESSEN, J. "Entre los Jibaros del Rio Zamora (Ecuador)," *Rivista Geografica Americana*, (1941), 16:367-74.

571 HO, Ting-jui. "Clothing and Ornaments Related to Atayal Head-hunting in the Department Collections." Taipei, Formosa, University, Department of Anthropology and Archaeology *Bulletin*, (1953), 2:22-29.

572 HODGE, F.W. "A Cheyenne Scalp." *Masterkey*, (1947), 21:97-99.

573 JEFFREYS, M.D. "Nsaangu's Head." *African Studies*, (1946), 5:57-62.

574 KARSTEN, Rafael. "The Head Trophy of the Jibaro Indians." PH, (1953), 520-30.

575 KELLY, W.H. "The Place of Scalps in Cocopa Warfare." *El Palacio*, (1949), 56:85-91.

576 KNOBLOCK, B. "Scalps." *Illinois State Archaeological Society, Quarterly Bulletin*, (1939), 2:6-7.

577 LEIGH, H. "Head Shrinking in Ancient Mexico." *Science of Man*, (1961), 2:4-7.

578 MASSARI, Claudia. "Considerazioni su di un raro cranio-trofeo dei mari del Sud." *Archivio per L'Anthropologia e La Ethnologia*, (1960), 90:279-87.

579 NADEAU, G. "Indian Scalping." *Bulletin of the History of Medicine*, (1941), X:178-94.

580 NADEAU, Gabriel. "Indian Scalping Technique in Different Tribes." *Ciba Symposia*, (1944), 5:1676-81.

581 NEUMANN, G. "Evidence for the Antiquity of Scalping from Central Illinois." *American Antiquity*, (1940), 5:287-89.

582 RITZENTHALER, R. "Shrunken Heads." *Lore* (Milwaukee), (1950), 3:88-91.

583 RYDEN, Stig. "Skalpierung bei den Tobaindianern," *Etnologiska studier*, (1935), (1):26-34. Goteborg.

584 SCHUSTER, Carl. "Head-hunting Symbolism on the Bronze Drums of the Ancient Dongson Culture and in the Modern Balkans." *4th Interna-*

tional Congress of Anthropological and Ethnological Sciences, (Vienna, 1952), 2:276-82.

585 SMITH, S.P. "Very Like Scalp-taking." JPS, (1919), 28:106.

586 TAYLOR, M.B. "A Jivaro War Trophy." BBSNS, (1915), XI, ii.

587 TURNER, G. "Counterfeit 'tsantsas' in the Pitt-Rivers Museum." *Man,* (1944), 44:57-58.

588 WOODBURY, George. "A Head Hunter's Trophy." *Colorado Magazine,* (1932), 9:214-17.

N / WAR CEREMONIES

589 AITCHISON, T.G. "Peace Ceremony as Performed by the Natives of the Ramu Headwaters." *Oceania,* (1936), 6:478-81.

590 ANDERSON, Frank. "The Marsai Celebrate the Royal Jubilee; Warrior Rites in Tanganyika Territory." *Illustrated London News,* (June 1935), 987.

591 ANDERSON, R. "North Cheyenne War Mothers." *Anthropology Quarterly,* (1956), 3:82-90.

592 AUFENANGER, H. "The War-Magic Houses in the Wahgi Valley and Adjacent Areas (New Guinea)." *Anthropos,* (1959), 54:1-26.

593 DENSMORE, Frances. "Sioux War Songs." AA, (1914), 16:122-23.

594 DORSEY, J.O. "Mourning and War Customs of the Kansas." *American Naturalist,* (1885), 19.

595 ——. "An Account of the War Customs of the Osages." *American Naturalist,* (1884), 18.

596 DOUGLASS, A.E. "A Find of Ceremonial Weapons in a Florida Mound." PAAAS, (1882), XXXI:585-92.

597 EVANS-PRITCHARD, E.E. "Nuer Spear Symbolism." *Anthropological Quarterly,* (1953), 26.

598 FRANK, Cedric Norman. *War with the Witch-doctor, an African Tale.* London, Universities' Mission to Central Africa, 1955.

599 FURER-HAIMENDORF, C. "The Headhunting Ceremonies of the Konyak Nagas of Assam." *Man,* (1938), 38:25.

600 GOLDSCHMIDT, W., G. Foster, and F. Essene. "War Stories from Two Enemy Tribes." JAFL, (1939), LII:141-54.

601 GUERNSEY, S.J. "Notes on a Navaho War Dance." AA, (1920), 22:304-7.

602 HAILE, B. *The Navajo War Dance.* St. Michael's, 1946.

603 HOEBEL, E. Adamson. "Song Duels among the Eskimo." In Bohannan, *Law and Warfare,* (1967), pp. 255-62.

604 HOERR, Charles L. "Drama in War." JAF, (1955), 68:253-81.

605 HOLSTI, R. "Some Superstitious Customs and Beliefs in Primitive Warfare." In *Festskrift tillegnad Edv. Westermarck.* Helsingfors, 1912.

606 KURK, R.F. "War Rituals at Zuni." *New Mexico,* (1945), XXIII, viii:14-15, 46-47.

607 LA FLESCHE, FR. "War Ceremonies and Peace Ceremony of the Osage Indians." *Smithsonian Institution Bulletin.* (1939), (101).

608 MATTHEWS, Robert H. "Message Sticks Used by the Aborigines of Australia." AA, (1908), 10:290.

609 McGUIRE, J.D. "Pipes and Smoking Customs of the American Aborigines." *Smithsonian Report, USNM.* Pt. 1, (1897), 361-645.

610 McKERN, W.C. "A Winnebago War-bundle Ceremony." YPMCM, (1928), VIII, i:146-55.

611 MOGGRIDG, L.T. "The Nyassaland Tribes, Their Customs and Their Poison Ordeal." JAI, (1912), 32:467-72.

612 MOONEY, J. "The Ghost-Dance Religion and the Sioux Outbreak." *Annual Report.* Bureau of American Ethnology, (1890), 2:653-1102.

613 MORIARTY, J.R. "Ritual Combat: A Comparison of the Aztec 'War of Flowers' and the Medieval 'Melee!' " *Colorado State College, Museum of Anthropology,* Miscellaneous Series, (1969), (9):1-22.

614 NORBECK, Edward. "Ritual Expression of Conflict, North American Indians and Subsaharan Africa." Paper presented at the 62nd annual meeting of the American Anthropological Association, San Francisco, 1963.

615 ———. "African Rituals of Conflict." AA, (1963), 65:1254-79.

616 OSTVOLD, T. "The War of the Aesir and Vanir: A Myth of the Fall of Nordic Religion." *Temenos: Studies in Comparative Religion,* (1969), 5:169-202.

617 PARSONS, Elsie C. *The Scalp Ceremonial of Zuni.* American Anthropological Association Memoir. No. 31, 1924.

618 ——. "Note on Navajo War Dance." AA, (1919), 21:465-67.

619 ——. "War God Shrines of Laguna and Zuni." AA, (1918), 20:381.

620 PERMAJAKOV, G. "Trubka mira. Istorija proishozdenija trubki mira i objcei, svjazanye s neju u severo-amerikanskih indejcer." Moscow, *Kokrug Sveta*, (1958), 9:44-45.

621 REAY, Marie. "Two Kinds of Ritual Conflict." *Oceania*, (1959), 29:290-96.

622 REINACH, A.J. "Les Trophees et les origines religieuses de la guerre." *Revue d'Ethnographie et de Sociologie*, (1913), 4:211-37.

623 ROBINSON, A.E. "The Carriage of Gods or Sacred Symbols in War." *Man*, (1934), 34:112.

624 SKINNER, Alanson. "War Customs of the Menomini Indians." AA, (1911), 13:299-312.

625 SPIER, Leslie. "The Sun Dance of the Plains Indians." *American Museum of Natural History, Anthropological Papers*, (1921) 16 (pt. 7).

626 TANNER, Nancy. "Minangkabau Disputes: The Use of Symbolic Products in Conflict Situations." Paper presented at the 67th annual meeting of the American Anthropological Association, Seattle, 1968.

627 TURLEY, F. "A Note On: War Dance." *American Indian Hobbyist*, (1960), 7:28-29.

628 WESTERMARCK, E. "Customs Connected with Homicide in Morocco." *Transactions of the Westermarck Society*, (1947), 1.

629 WILDSCHUT, William. *Crow Indian Medicine Bundles*. New York: Museum of the American Indian, Heye Foundation, 1960, pp. 39-94.

630 ——. "Crow War Bundle of Two-Leggings." *Indian Notes*, (1926), 3:284-88.

O / MILITARY ORGANIZATION, DEFENSE, COSTUMES

631 "The Measurement and Utilization of Brain Power in the Army." *Science*, (1919), 49 (1262):221-26; 49 (1263):251-59.

632 "Palefaces' Defense System Old Stuff to Indians." *The Totem Pole*, (1943), 5:4-5.

633 "Tactics and Techniques of Cavalry." *Cavalry Journal*. 6th ed. Washington, 1935.

634 ABEL, Theodore. "The Element of Decision in the Pattern of War." ASR, (1941), 6:853-59.

635 ANDRZEJEWSKI, Stanislaus. *Military Organization and Society*. Foreword by A.R. Radcliffe-Brown. London: Routledge and Kegan Paul, 1954.

636 ANTHROPOVA, V.V. "Problems of War Organization and Warfare Among the Peoples of the Far North-East Siberia, 17th-18th Century." (in Russian). Moscow: *Trudy Instituta Etnografii Imena N.N. Mikluho-Maklaja, Novaja Serija*, 1957.

637 BAJWA, F.S., and M.B. Dass. *Military System of the Sikhs: During the Period 1799-1849*. Delhi, 1964.

638 BEARSLEY, H.G. "Short Notes on Two Maori Pa in English Museums." JPS, (1932), 41:237-38.

639 BEEMER, H. "The Development of the Military Organization in Swaziland." *Africa*, (1937), 10:55.

640 BENSON, Michael G. "Age Grade and Military Organization: Another View of Predatory Expansion." Paper presented at the 49th annual meeting of the Central States Anthropological Society, Milwaukee, 1969.

641 BEST, E. *The Pa Maori: An Account of the Fortified Villages of the Maori in Pre-European and Modern Times*. Wellington: Whitcombe and Tombs, 1927.

642 ———. "Stockades and Earthworks in New Zealand." *American Antiquarian*, (1895), 17:154-56.

643 BORGONIO, Gaspar G. "Organizacion militar de los Tenochca." *Revista Mexicana de Estudios Anthropologios*, (1954-55), 14:361-83.

644 BOTTEN, W. "An Indian Drill." *Smoke Signals*, (1961), 4:6.

645 BOYD, R.K. "How the Indians Fought: A New Era in Skirmish Fighting. By a Survivor of the Battle of Birch Cooley." *Minnesota History*, (1930), 11:299-304.

646 BUCHLER, Ira R., and H.A. Selby. *Kinship and Social Organization*. New York: Macmillan Company, 1968. (Chps. 5 and 6 on "Alliance Theory.")

647 BURNE, Alfred H. *The Art of War on Land*. Harrisburg, Pennsylvania: Military Service Publishing Co., 1941 (2nd ed. London: Methuen, 1950).

648 CHRISTOPHE, L.A. "L'Organisation de l'armée égyptienne à l'epoque Ramesside." Cairo: *Revue du Caire*, (1957), 20:387-405.

649 DECARY, R. *Coutumes guerrières et organisation militaire chez les anciens Malgaches.* v. 1, *Les Anciennes pratiques de guerre.* v. 2, *L'Histoire militaire des Merina.* Paris: Maritimes et d'Outremer, 1966.

650 Du CHOUL, Guillaume. *Découvre sur la castramentation et discipline militaire des anciens Romains, des bains and antiques exercitations grecques and romaines.* Lyon: Guillaume Roville, 1881.

651 DYER, D.B. *Fort Reno or Picturesque Cheyenne-Arapaho Army Life before the Opening of Oklahoma.* New York: G.W. Dillingham, 1896.

652 EBISAWA, T. "Gencho Tanmagun Kenkyu josetsu." (Study of Tanma-army of the Mongol Empire.) *Shiryu*, (1966), 7:50-65.

653 ELRY, W.J. "Te Rae-o-Te-Karaka. A Pa or Fortified Village in Queen Charlotte Sound." JPS, (1927), 36:367-68.

654 EWERS, J.C. "The Blackfoot War Lodge: Its Construction and Use." AA, (1944), 46:182-92.

655 FARRER, J.A. *Military Manners and Customs.* London: H. Holt & Co., 1885.

656 FAULKNER, R.O. "Egyptian Military Organization." JEA, (1953), 39:32-47.

657 FERGUSON, W.S. "The Zulus and the Spartans: A Comparison of their Military Systems." *Varia Africana*, v. 2. Harvard African Series. Cambridge: Harvard University Press, 1918.

658 FIRTH, Raymond. "Maori Hill-forts." *Antiquity*, (1927), 1:66-78.

659 FOSBROOKE, H.A. "The Defensive Measures of Certain Tribes in North-Eastern Tanganyika: Mbugwe Flats and Sonjo Scarps." TNR, (1955), 39:1-11.

660 ———. "The Defensive Measures of Certain Tribes in North-Eastern Tanganyika: Chagga Forts and Bolt Holes." TNR, (1954), 37:115-129.

661 ———. "The Defensive Measures of Certain Tribes in North-Eastern Tanganyika: Iraqw Housing as Affected by Inter-tribal Raiding." TNR, (1954), 36:50-51.

662 FRIED, M.H. "Warfare, Military Organization and the Evolution of Society." *Anthropologica*, (Ottawa, 1961), 3:134-47.

663 FULLER, J.F.C. "Tactical Formations." *Encyclopedia Britannica* 14th ed., (1929), 21:739.

664 HALL, A.R. "A Note on Military Pyrotechnics." In C. Singer, ed., *A History of Technology.* Oxford: University Press, (1956), v. 2, pp. 374-82.

665 HARMON, E.M. "The Story of the Indian Fort near Granby, Colorado." *Colorado Magazine,* (1945), 22:167-71.

666 HEYDEN de ALVAREZ, D. "Costume of the Warrior of Tenochtitlan." *Science of Man,* (1960), 1:12-16.

667 KELLOG, L.P. "The Stockaded Village." *Wisconsin Archeologist,* (1929), 8:61-68.

668 KELLY, L.G. "Mangatoatoa Pa." JPS, (1933), 42:167-78.

669 KOSVIN, M.O. "Kvoprusu o voennoj demokratii." Moscow: *Trudy Instituta Etnografii im N.N. Mikluho-Maklaja. Novaja Serija,* (1960), 54:241-61.

670 KUBBEL, L.E. "O nekoforyh certah voennoj sistemi Kalifata Omajjadov (661-750 A.D.)." (Omayed Caliphate War System. Has English summary.) Leningrad, *Palestinskij Sbornik,* (1959), 4:112-32.

671 LEWIS, D.G. "Lobengula's Regiments: Recruiting and Lobolo." Salisbury, *Nada,* (1956), 33:5.

672 LOWIE, R.H. "Military Societies of the Crow Indians." APAM, (1913), XI:143-217.

673 ———. "Property Rights and Coercive Powers of Plains Indian Military Societies." JLPS (1943), 1 (3-4):59-71.

674 MATSUOKA, Asa. "Battle Dress of Feudal Japan." *Asia,* (1932), 32:290-97.

675 McCLINTOCK, W. "Blackfoot Warrior Societies." M, (1937-38), XI:148-58, 198-204; XII:11-23.

676 MELMAN, Seymour. "Decision Making on War and Peace." WAACA, (1968), 229-34.

677 MILLOT, J. "Review of Decary, 1966, Coutumes guerrieres. . . . " Paris, *Objets et Mondes,* (1966), 6:342.

678 OTTERBEIN, Keith. "Cross-cultural Studies of Armed Combat." Studies in International Conflict, Research Monograph No. 1, *Buffalo Studies,* (1968), 4:91-109.

679 PALERM, A. "Notas sobre las construcciones militares y la guerra en Mesoamerica." Mexico, *Anales de Instituto Nacional de Antropologia e Historia*, (1956), 8:123-36.

680 PETERSEN, Karen. "On Hayden's List of Cheyenne Military Societies." AA, (1965), 67:469-72.

681 ———. "Cheyenne Soldier Societies." *Plains Anthropologist*, (1944), 9(9).

682 RAHARIJAONA, Dr. "Le Mode de recrutement des soldats par la reine." Tananarive, *Bulletin de L'Academie Malgache*, (1957), 34:105-6.

683 REVON, M. "Les Coutumes de la guerre dans le Japan primitif." *L'Ethnographie*, (1931), 23:51-55.

684 ROSENFELD, H. "The Social Composition of the Military in the Process of State Formation in the Arabian Desert." JAI, (1965), 95:75-86.

685 SCHULMAN, A.R. *Military Rank, Title and Organization in the Egyptian New Kingdom*. Berlin: Hessling, 1964.

686 SECOY, Frank Raymond. "Changing Military Patterns on the Great Plains." *Monograph of the American Ethnological Society*, (1953), (21).

687 SKINNER, W.H. "The Ancient Fortified Pa." JPS, (1911), 20:71-77.

688 SPICER, E.H. "The Military Orientation in Yaqui Culture." FTD, (1950), 171-88.

689 SUKRU, Tahsin. "Bir harp plani." (A Military Plan.) *Turk Tarih, Arkeologya ve etnografya dergisi*, (1934), 2:254-57.

690 SUMMERS, R.F. "The Military Doctrine of the Matebele." *Nada*, (1955), 32:7-15.

691 THORNDIKE, E.L. "Scientific Personnel Work in the Army." *Science*, (1917), 49:53-61.

692 TOY, Sidney. *A History of Fortification from 3000 B.C.-A.D. 1700*. London: Heinemann, 1955.

693 URAY, G. "The Four Horns of Tibet according to the Royal Annals." Budapest, *Acta Orientalia*, (1960), 10:31-57.

694 URBAIN-FAUBLEE, M. "Review of Decary, 1966, *Coutumes guerrières. . . .*" London, *Africa*, (1967), 37:369-70.

695 VALETTE, J. "Review of Decary, 1966, *Coutumes guerrières. . . .*" Tananarive, *Bulletin de Madagascar,* (1966), 244:902-04.

696 WHITE, Thain. *The Battle Pitts of the "Koyokees."* Missoula, Montana State University, Anthropology and Sociology Papers, No. 10, 1952.

P / WEAPONS: GENERAL

697 "Red Man Used Poison Gas Warfare." *El Palacio,* (1932), 32:173-74.

698 "The Weapon." *The Living Museum,* (1945), 6:83.

699 "Weapons and Culture." *New World Antiquity,* (1959), 6:80-87.

700 BEACHEY, R.W. "The Arms Trade in East Africa in the Late Nineteenth Century." *Journal of African History,* (1962), 3:451-67.

701 BEAUCLAIR, I. "Fightings and Weapons of the Yami of Botel Tobago." (Melanesia). Taipei, *Bulletin of the Institute of Ethnology, Academia Sinica,* (1958), 5:87-111.

702 BOE, John. "Armateurs en os préhistoriques et leurs paralleles ethno-graphiques." *L'Anthropologie,* (1935), 45:591-600.

703 BRETON, A.C. "Arms and Accoutrements of the Ancient Warriors at Chichen Itza." AA, (1909), 11:456-67.

704 CHILDE, V.G. "Horses, Chariots and Battle-axes." *Antiquity,* (1941), 15:196-99.

705 CLARKE, H. "On Prehistoric Names of Weapons." JAI, (1877), 6:142-49.

706 COWPER, H.S. *The Art of Attack. Being a Study in the Development of Weapons and Appliances of Offense, from the Earliest Times to the Age of Gunpowder.* Ulverston, England: Holmes, 1906.

707 CURTIS, F.S. "Spanish Arms and Armor in the Southwest." NMHR, (1927), 2:107-33.

708 ———. "The Influence of Weapons on New Mexico History." NMHR, (1926), 1:324-34.

709 DEMMIN, Auguste. *Weapons of War: Being a History of Arms and Armour.* London: Bell and Daldy, 1870.

710 DERANIYAGALA, P. "Sinhala Weapons and Armor." *Journal of the Ceylon Branch of the Royal Asiatic Society,* (1942), 95:97-142.

711 EGERTON, Wilbraham. *An Illustrated Handbook of Indian Arms.* London: William Allen, 1880.

712 FALKINDER, J.S. "Throwing Stones." *Mankind*, (1932), 1:90-91.

713 FULLER, J.F.C. *Armament and History. A Study of the Influence of Armament on History from the Dawn of Classical Warfare to the Second World War.* London: Eyre & Spottiswoode, 1946.

714 GALIMBERTI MIRANDA, C-A. "Las armas de Guerra Incaicas i su evolucion." Cuzco Universidad, *Instituto y Museo Arqueologico, Revista*, (1951), 13-14:87-137.

715 GARDNER, G.B. *Keris and Other Malay Weapons.* Singapore: Progressive Publishing Co., 1936.

716 GORDON, D.H. "Fire and the Sword: The Technique of Destruction." *Antiquity*, (1953), 27:149-52.

717 GRIESBACH, C.L. "Weapons and Implements Used by Kaffin Tribes and Bushmen of South Africa." JAI, (1881), 1:154-55.

718 HADDON, A.C. "Weapons and Objects Used in Warfare." RCETS, (1912), 4:172-204.

719 HALL, A.R. "Military Technology." In C. Singer, ed., *A History of Technology.* Oxford: University Press, (1956), v. 2, pp. 695-730.

720 HAMBLY, W.D. "The Preservation of Local Types of Weapons and Other Objects in Western Australia." AA, (1931), 33:1-15.

721 HARVARD UNIVERSITY. Peabody Museum of Archaeology and Ethnology. Subject Catalogue of the Library, v. 26, "Technology-Weapons," pp. 361-93. (Over 800 sources are listed.) Boston: G.K. Hall, 1963.

722 HEIZER, R.F. "The Use of 'Poison Gases' in Warfare by the American Indians." *Ciba Symposia*, (1944), 6:1906-7.

723 HUGHES, A.J., and R. Summers. "The Matabele Warrior; His Arms and Accoutrements." Cambridge, England. *Bulawayo*, National Museums of Southern Rhodesia, Occasional Papers, (1955), 2:779-91.

724 HUTTON, J.H. "The Bolas and Its Distribution." *Man*, (1943), 48:96.

725 KNIGHT, E.H. "A Study of the Savage Weapons at the Centennial Exhibition, Philadelphia." Washington, *Smithsonian Institution, Annual Report for 1879*, (1886), 213-97.

726 KREBS, W. "Blefanten in den Heeren der Antike." Rostock, *Wissenschaftliche Zeitschrift der Universität Rostock. Gesellschafts und Sprachwissenschaftliche Reihe*, (1964), 13:205-220.

727 KRIEGER, H.W. *The Collection of Primitive Weapons and Armor at the Philippine Islands in the United States National Museum.* Washington: Government Printing Office, 1926.

728 LEGASSICK, M. "Firearms, Horses and Samorian Army Organization, 1870-1898." *Journal of African History*, (1966), 7:95-115.

729 MacKENZIE, A. "Descriptive Notes on Certain Implements, Weapons, etc. from Graham Island." PTRSC, (1891), IX, ii:45-59.

730 METSCHL, John. *The Rudolph J. Nunnemacher Collection of Projectile Arms.* 2 vols. Wisconsin: Milwaukee Public Museum, Bulletin, No. 9, 1928.

731 MONTAGU, Ashley. "Predators, Tools, Implements, and Weapons: A Comment." AA, (1969), 71:312-13.

732 MONTAGUE, Col. L.A.D. *Weapons and Implements of Savage Races (Australia, Asia, Oceania & Africa).* London: 1921.

733 MURDOCH, G.M. "Gilbert Islands, Weapons and Armour." JPS, (1923), 22:174-75.

734 NORDENSKIOLD, E. "Palisades and 'Noxious Gases' among the South-American Indians." *Ymer*, (1918), 220-43.

735 OAKESHOTT, R. Ewart. *The Archeology of Weapons. Arms and Armor from Prehistory to the Age of Chivalry.* New York: Praeger, 1960.

736 PAYNE-GALLWEY, R. *Projectile-throwing Engines.* London: 1917.

737 PETRIE, W.M.F. *Tools and Weapons* (of Ancient Egypt). London: 1917.

738 RUSSEL, Carl P. *Guns on the Early Frontier.* Berkeley: University of California Press, 1957.

739 SALONEN, A. *Notes on Waggons and Chariots in Ancient Mesopotamia.* Helsinki: 1950.

740 SAPPER, Karl. "Central American Weapons in Modern Use." *Globus*, (1903), v. 83, No. 4.

741 SCHAFER, E.H. "War Elephants in Ancient and Medieval China." Leiden, *Oriens*, (1957), 10:289-91.

742 SCHULLER, R.R. "On the Supposed Use of Poison by the Xinca Indians of Guatemala and the Pipil of Cuzcatan." *Indian Notes*, (1930), 7:153-63.

743 SELIGMANN, C.G. "Note on the Preparation and Use of the Kenyah Dartpoison Ipoh." JAI, (1902), 32:239-44.

744 SMITH, S.C. "An Analysis of the Firearms and Related Specimens from Like-A-Fishook Village and Fort Berthold I." PIA, (1955), IV:3-12.

745 SMITH, T.H. "Maori Implements and Weapons." *Transactions and Proceedings at the New Zealand Institute*, (1893), 26:423-52.

746 SNOWDEN, A. "Some Technological Notes on Weapons and Implements used in Mashonaland." *Nada*, (1940), 17:62-70.

747 SPEISER, F. "Über Schutzwaffen in Melanesien." *Internationales Archiv für Ethnographie*, (1941), 40:81-121.

748 THAYER, B.W. "Features Distinguishing the Hudson's Bay Musket." *Minnesota Archaeologist*, (1947), 12:34-35.

749 WORCHESTER, D.E. "The Weapons of American Indians." NMHR, (1945), 20:227-38.

750 YADIN, Y. "Hyksos Fortifications and the Battering-Ram." BASOR, (1955), 137:23-32.

P.1 / Armor & Shields

751 AUFENANGER, H. "The Parry Shield in the Western Highlands of New Guinea." *Anthropos*, (1957), 52:631-33.

752 BURGESS, E. "Further Research into the Construction of Mail Garments." *Antiquaries Journal*, (1953), 33:193-202.

753 CURTIS, F.S., Jr. "Spanish Arms and Armor in the Southwest." *New Mexico Historical Review*, (1927), 2:107-33.

754 GUDGER, E.W. "Helmets from Skins of the Porcupine-fish." *Scientific Monthly*, (1930), 30:432-42.

755 HALL, H.U. "Some Shields of the Plains and Southwest." *Pennsylvania University Museum Journal*, (1926), 17:37-61.

756 HENCKEN, Hugh O. *Helmets of the European Bronze Age, with Some of the Iron Age*. Cambridge: Harvard Peabody Museum, 1971.

757 HODGE, F.W. "Ceremonial Shields of Taos." *El Palacio*, (1926), 20:231-34.

758 HOUGH, W. "Primitive American Armor." Smithsonian Institution, *Report of the U.S. National Museum, 1893*, (1895), 625-51.

759 LAMBERT, M.F. "Oldest Armor in the United States Discovered at San Gabriel del Yungre." *El Palacio*, (1952), 59:83-87.

760 LHOTE, H. "Note sur l'origine des lames d'epée des Touavega." *Notes Africaines*, (1954), 61:9-12.

761 LORIMER, H.L. "Defensive Armor in Homer, with a Note on Women's Dress." *Annals of Archaeology and Anthropology,* (1928), 15:89-129.

762 PHILLIPPS, W.J. "Breast Plates of Fiji." *Fiji Society of Science and Industry Transactions and Proceedings,* (1953), 4:52-53.

763 RYAN, D.J. "Some Decorated Fighting-Shields from the Mendi Valley, Southern Highlands, District of Papua." *Mankind,* (1958), 5:243-49.

764 SCHEBESTA, P. "Der afrikanische Schild." *Anthropos,* (1924), 18-19; 1012-62.

765 SELER, E. "Ancient Mexican Shields." *Internationales Archiv für Ethnographie,* (1892), 5:168-72.

766 SPEC, F.G. "A Plains Indian Shield and Its Interpretation." *Primitive Man,* (1948), 21:74-79.

767 THORDEMAN, Bengr. "The Asiatic Splint Armour in Europe." *Acta Archaeologica,* (1933), 4:117-50.

768 TYLDEN, G. "Bantu Shields." *South African Archaeological Bulletin,* (1946), 2:33-37.

769 WARDLE, H. "Defensive and Offensive Power of the Shield." (Dakota Sioux) *University of Pennsylvania Museum Bulletin,* (1938), 2:24-27.

770 WILDSCHUT, William. "A Crow Shield." *Indian Notes,* (1925), 2:315-20.

P.2 / Axes, Clubs, & Swords

771 "Sioux Indian War Clubs." Washington, *Museum News,* (1948), 9 (8):1-4.

772 "A Viking Sword." *Royal Ontario Museum of Archaeology,* (1929), 8:12-13.

773 BAILEY, B.A. "Notes on Oceanian War Clubs." JPS, (1947), 56:3-17.

774 BATES, O. "On the Origin of the Double-bladed Swords of the West Coast." *Harvard African Studies,* (1918), 2:187-93.

775 BENTON, S. "An Unlucky Sword: The Leaf-shaped Blade from Mycenae." *Man,* (1931), 31:127-28.

776 BRASSER, T. "War Clubs." *American Indian Tradition,* (1961), 3:77-83.

777 BURKE, R.P. "Pipe-Tomahawks." *Arrow Points,* (1931), 19:5-6.

778 CASSON, Stanley. "Battle-axes from Troy." *Antiquity,* (1933), 7:337-39.

779 CHURCHILL, W. *Club Types of Nuclear Polynesia.* Washington: Carnegie Institution of Washington, Publication No. 255, 1917.

780 COGHLAN, H. "The Evolution of the Axe from Prehistoric to Roman Times." JAI, (1943), 73:27-56.

781 DAVIDSON, D.S. "Stone Axes of Western Australia." AA, (1938), 40:38-48.

782 DERRICK, R.A. "Notes on Fijian War Clubs, with a System of Classification." JPS, (1957), 66:391-95.

783 GORDON, Col. D.H. "Swords, Rapiers and Horse-riders." *Antiquity,* (1953), 27:67-78.

784 GRANCSAY, S.V. "Three Primitive Japanese Swords." *Bulletin of the Metropolitan Museum of Art,* (1932), 27:208-10.

785 GROTTANELLI, V.L. "On the 'Mysterious' Baratu Clubs from Central New Guinea." *Man,* (1951), 51:105-7.

786 HADDON, A.C. "A Classification of the Stone Clubs of British New Guinea." JAI, (1910), 30:221-50.

787 HARRISON, T.H. "Stone Weapons from Borneo." JPS, (1950), 59:169.

788 HODGE, F.W. "Kwakiutl Sword." *Indian Notes,* (1924), 1:200-4.

789 HOLMES, William H. "The Tomahawk." AA, (1908), 10:264-76.

790 HOUGH, Walter. "The Corrugation in African Sword Blades and Other Weapons." Washington, *U.S. National Museum Bulletin,* (October 1888), 172.

791 IVENS, W.G. "Clubs Called Lwari-i-hua." *Ethnologia Cranmorensis,* (1938), 2:8-18.

792 JEFFREYS, M.D. "Ibo Club Heads." *Man,* (1957), 67:57-58.

793 KENYON, K.M. "A Crescentic Axe-head from Jericho, and a Group of Weapons from Tell el Hesi." *Eleventh Annual Report of the Institute of Archeology, University of London,* 1955.

794 LECHLER, G. "The Beginning of the Bronze Age and the Halherd." *American School of Prehistoric Research, Bulletin,* (1938), 14:3-53.

795 NEEDLER, W. "An Egyptian Battle-Ax." *Archaeology,* (1952), 5:48-50.

796 SAVILLE, Marshall H. *Bladed Warclubs from British Guiana.* New York, Museum of the American Indian, Indian Notes and Monographs, No. 14, 1921.

797 SCHELLBACH, L. "An Historic Iroquois Warclub." *Indian Notes*, (1928), 5:157-66.

798 SHELFORD, R. "A Provisional Classification of the Swords of the Sarawak Tribes." *Journal of the Anthropological Institute*, (1901), 4:219-28.

799 SKINNER, Alanson. "An Old Seneca Warclub." *Indian Notes*, (1926), 3:45-47.

800 SKINNER, H.D. "Weapons of Coconut Wood from the Cook Islands." JPS, (1943), 52:86-89.

801 VULLIAMY, C.E. "Discovery of a Saxon Sword in Wales." *Man*, (1931), 31:86-87.

802 WICKERSHAM, J. "An Aboriginal War Club." *American Antiquarian*, (1895), 2:72-77.

803 WIDSTRAND, Carl G. "African Axes." Uppsala, *Studia Ethnographica Upsaliensia*, (1958), 15:1-164.

804 WOODWARD, A. "The Meta Tomahawk; Its Evolution and Distribution in North America." *Fort Ticonderoga Museum Bulletin*, (1946), 3:2-42.

P.3 / Bows, Arrows, & Crossbows

805 "Archery in China." *Field Museum News*, (1938), 6:3.

806 ALBRIGHT, W.F., and G.E. Mendenhall. "The Creation of the Composite Bow in Canaanite Mythology." JNES, (1942).

807 ALM, Josef. "Bows and Bow-Shooting among the Lapps." *Ethnos*, (1936), 1:153-60.

808 BACON, Raymond F. "Philippine Arrow Poisons." *Philippine Journal of Science*, (1905), III, No. 1. (reviewed in AA, (1908), 10:357.)

809 BALFOUR, H. "Notes on the Composite Bow from Hunza." *Man*, (1932), 32:161.

810 ——. "Method Employed by Natives of Australia in the Manufacture of Glass Spear Heads." *Man*, (1903), 3:65.

811 ——. "On the Structure and Affinities of the Composite Bow." *Journal of the Anthropological Institute*, (1889), 19:225-50.

812 ———. "On a Remarkable Ancient Bow and Arrows Believed to Be of Assyrian Origin." JAI, (1897), 26:210-20.

813 ———. "The Origin of West African Crossbows." *Journal of African Society*, (1909), 8:337-56.

814 BOONEY, A. "The American Indian as an Archer." *Smoke Signals*, (1955), 4:2-3.

815 BRUES, A. "The Spearman and the Archer—An Essay on Selection in Body Build." AA, (1959), 61:457-69.

816 BUSHNELL, D.I., Jr. "The Bows and Arrows of the Arawak in 1803." M, (1910), X:22-24.

817 CARPENTER, E.S., and R.B. Hassrick. "Some Notes on Arrow Poisoning among the Tribes of the Eastern Woodlands." *Proceedings of the Delaware County Institute of Science*, (1947), X, 45-52.

818 CODRINGTON, R.H. "On Poisoned Arrows in Melanesia." JAI, (1890), 19:215-19.

819 CORDY, N. "A Possible Origin for the Bow." *Masterkey*, (1961), 35:150-51.

820 CURRELLY, C.T. "The Crossbow." *Royal Ontario Museum of Archaeology Bulletin*, (1931), 10:11-12.

821 CUSHING, F.H. "The Arrow." AA, (1895), 8:307-49.

822 ELLIS, F.H. "Laguna Bows and Arrows." *El Palacio*, (1959), 66:91.

823 ELLSWORTH, C. "Bows and Arrows." *Masterkey*, (1950), 24:5-14, 56-66.

824 ELMER, Robert P., ed. *The Book of the Long Bow*. Garden City, New York: Doubleday, Doran, 1929.

825 EMENEAU, M.B. "The Composite Bow in India." *American Philosophical Society, Proceedings*, (1953), 97:77-87.

826 EVANS, O.F. "The Development of the Atlatl and the Bow." *Texas Archeological Society Bulletin*, (1961), 30:159-62.

827 FARIS, N.A., and R.P. Elmer. *Arab Archery*. Princeton: Princeton University Press, 1945.

828 FISCHER, H.G. "The Archer as Represented in the First Intermediate Period." JNES, (1966), 21:50-52.

829 FLINT, W. "The Arrow in Modern Archery." AA, (1891), 4:63-67.

830 GOODWIN, A.J. "Some Historical Bushman Arrows." *South African Journal of Science*, (1945), 41:429-43.

831 HOLMES, W.H. "Manufacture of Stone Arrow Points." AA, (1891), 4:49-58.

832 HORNELL, J. "South Indian Blowguns, Boomerangs, and Crossbows." JAI, (1924), 54:316-46.

833 HOUGH, W. "Arrow Feathering and Pointing." AA, (1891), 4:60-63.

834 HOWARD, J.H. "Notes on Dakota Archery." University of South Dakota Museum, *Museum News*, (1950), 2:1-3.

835 HUTTON, J.H. "The Use of the Bow among the Naga Tribes of Assam." *Folk-Lore*, (1922), 33:305-6.

836 JACKSON, A.T. "Indian Arrow and Lance Wounds." *Texas Archeological and Paleontological Society, Bulletin*, (1943), 15:38-65.

837 KARSTEN, R. "Addenda to My Notes on South American Arrow-poison." CHL, (1934), VI, viii.

838 ———. "Notes on South American Arrow-poison." CHL, (1933), VI, iv.

839 KNOWLES, W.J. "On the Classification of Arrow Heads." JAI, (1877), 6:482-84.

840 KROEBER, A.L. "Arrow Release Distributions." *University of California, Publications in American Archaeology and Ethnology*, (1927), 23:283-96.

841 La CHARD, L.W. "The Arrow-Poisons of Northern Nigeria." *Journal of the African Society*, (1905), 5:122-27.

842 LANE, F. "The Pellet Bow among South American Indians." ICA, (1955), XXXI, i:257-66.

843 LEAKEY, L.S.B. "A New Classification of the Bow and Arrow in Africa." JAI, (1926), 56:259-99.

844 LONGMAN, C.J. "The Bows of the Ancient Assyrians and Egyptians." JAI, (1894), 24:49-57.

845 LYONS, A.P. "The Arrows of the Upper Morehead River (Papua) Bush Tribes." *Man*, (1922), 22:145-74.

846 MASON, O.T. "North American Bows, Arrows and Quivers." ARSI, (1893), 631-79.

847 ——, et al. "Arrow and Arrow-makers." AA, (1891), 4:45-74.

848 McDANIEL, W.B. "The So-Called Bow-puller of Antiquity." *American Journal of Archaeology*, (1918), 22:25-43.

849 McLEOD, W.E. "An Unpublished Egyptian Composite Bow in the Brooklyn Museum." *American Journal of Archaeology*, (1958), 62:397-401.

850 MEYER, H. "Bows and Arrows in Central Brazil." Washington, *Smithsonian Institution Annual Report for 1896*, (1898), 549-82.

851 MORSE, E.S. *Additional Notes on Arrow-Release.* Salem, Massachusetts: Peabody Museum, 1922.

852 ——. "Ancient and Modern Methods of Arrow-release." *Essex Institute Bulletin*, (October-December 1885).

853 MURDOCH, J. "Study of Eskimo Bows in the U.S. National Museum." RUSNM, (1884), ii:207-16.

854 ODENDAAL, P.J. "The Bow and Arrow in Southern Rhodesia." *Nada*, (1930), 8:59-61; (1941), 18:23-24.

855 PEET, S. "The Boomerang and the Bow and Arrow." *American Antiquarian*, (1905), 27:233-40.

856 PHILLIPPS, W.J. "The Use of the Bow and Arrow in New Zealand." *Ethnos*, (1954), 19:139-42.

857 POND, A.W. "Fashioning of Arrowheads by Primitive Methods." *El Palacio*, (1930), 28:181-82.

858 POPE, S.T. "A Study of Bows and Arrows." *University of California, Publications in American Archaeology and Ethnology*, (1923), 13:329-414.

859 PRINS, A. "A Teita Bow and Arrows." *Man*, (1955), 55:33-35.

860 ROGERS, S.L. "The Aboriginal Bow and Arrow of North America and Eastern Asia." AA, (1940), 42:255-69.

861 SANTESSON, C.G. "An Arrow Poison with Cardiac Effect from the New World." CES, (1931), IX:155-87.

862 SCHAPERA, I. "Bushman Arrow Poisons." *Bantu Studies*, (1925), 2:199-214.

863 SCHEBESTA, P. "The Bow and Arrow of the Semang." *Man*, (1926), 26:88-89.

864 SHARP, B. "On Bows and Arrows and Other Implements Found among the Arctic Highlanders." *Proceedings of the Academy of Natural Sciences of Philadelphia*, (1891), 451-54.

865 TANG, Mei-Chun. "A Comparative Study of Bows and Arrows of the Formosan Aborigines." *Ethnological Society of China, Bulletin*, (1955), 1:139-70.

866 WABLA, E.J. "Indian Bows and Arrows." *Totem Pole*, (1948), 6:1-4.

867 WILLIAMS, L. "Osage Orange Wood Prized by Indians for Bows." *Field Museum News*, (1935), 8:3.

868 WILSON, T. "Arrowpoints, Spearheads and Knives of Prehistoric Times." Washington, *Report of the U.S. National Museum for 1897*, 811-988.

869 YADIN, Y. "The Composite Bow of the Canaanite Goddess Arath." BASOR, (1947), 107:11-15.

P.4 / Blowguns, Boomerangs, & Slings

870 BOGLAR, L. "Some More Data on the Spreading of the Blowgun in South America." *Acta Ethnographica*, (1950), 1:121-37.

871 BUNKERAH, Tayo. "The Sling in New Caledonia." *Mankind*, (1934), 1:190.

872 CHILDE, V.G. "The Significance of the Sling for Greek Pre-history." In *Studies Presented to David Moore Robinson*. St. Louis, 1951.

873 CHINNERY, E.W.P. "The Blowgun in New Britain." *Man*, (1927), 27:208.

874 CORNISH, J.J. "The Mystery of the Boomerang." NH, (1956), 65:242-45.

875 FOX-PITT-RIVERS, A.H.L. "On the Egyptian Boomerang and its Affinities." JAI, (1883), 12:454-63.

876 GILL, R.C. "Curari, the Flying Death; The Primitive and Mysterious Poison Used by Natives of the Amazon Jungles in Their Blowguns May Prove of Value to Civilized Medical Science." NH, (1935), 36:378-92.

877 GULART, J. "Les Boomerangs d'Australie." *La Revue de Géographie Humaine et d'Ethnologie*, (1949), 4:25-34.

878 HALL, J.C. "A Pharmaco-Bacteriologic Study of Two Malayan Blow-gun Poisoned Darts." AA, (1928), 30:47-59.

879 HORNELL, J. "South Indian Blowguns, Boomerangs, and Crossbows." JAI, (1924), 54:316-46.

880 KRAUSE, F. "Sling Contrivances for Projectile Weapons." Washington, Smithsonian Institution, *Annual Report* (1904), 619-38.

881 LINNE, Sigvald. "Blow-guns in Ancient Mexico." *Ethnos,* (1939), 4:61-65.

882 MEAD, C.W. "The South American Blow-gun." AMJ, (1908), VIII:42-43.

883 MEANS, P.A. "Distributions and Use of Slings in Pre-Columbian America." Washington, *U.S. National Museum, Proceedings,* (1920), 55:317-49.

884 NASH, C.H. "Choctaw Blowguns." *Tennessee Archaeologist,* (1960), 16:1-19.

885 NIES, J.B. "The Boomerang in Ancient Babylonia." AA, (1914), 16:26-32.

886 PEET, S. "The Boomerang and the Bow and Arrow." *American Antiquarian,* (1905), 27:233-40.

887 RILEY, Carroll. "Early Accounts of the South and Central American Blowgun." Boulder, *University of Colorado Studies in Anthropology,* (1954), 4:78-89.

888 SPRINTZIN, N.G. "The Blowgun in America, Indonesia, and Oceania." New York, *International Congress of Americanists, 23rd Session, Proceedings, 1928,* (1930), 699-704.

889 TURCK, A. *Théorie, Fabrication et Lancement des Boomerangs.* Paris: Éditions Chiron, 1952.

890 YDE, Jens. "The Regional Distribution of South American Blowgun Types." *Societé des Américanistes, Journal,* (1948), 37:275-317.

P.5 / Daggers, Spears, & Throwing Sticks

891 BRYAN, F. "A Choctaw Throwing Club." *Masterkey,* (1933), 7:178-79.

892 BUSHNELL, G.H.S. "Some Old Western Eskimo Spearthrowers." *Man,* (1949), XLIX:121.

893 CURWEN, E.C. "Spear-throwing with a Cord." *Man,* (1934), 34:105-6.

894 DAVIDSON, D.S. "Australian Throwing-sticks, Throwing-clubs, and Boomerangs." AA, (1936), 38:76-100.

895 ———. "Australian Spear-traits and Their Derivations." JPS, (1934), 43:41-72, 142-62.

896 EVANS, O. "How to Use the Atlatl." *Oklahoma Anthropological Society, Newsletter*, (1957), 6 (9).

897 HEIZER, R.F. "Introduced Spearthrowers (Atlatls) in California." M, (1945), XIX:109-12.

898 HILL, M.W. "The Atlatl or Throwing Stick." *Tennessee Archaeologist*, (1948), 4:37-44.

899 MASON, J.A. "Some Unusual Spearthrowers of Ancient America." *Pennsylvania University Museum Journal*, (1928), 19:290-334.

900 MASSEY, W.C. "The Survival of the Dart-thrower on the Peninsula of Baja California." SWJA, (1961), 17:81-93.

901 MILIK, J.T., and F.M. Cross. "Inscribed Javelin-Heads from the Period of the Judges." BASOR, (1954), 134:5-15.

902 MURDOCH, J. "This History of the 'Throwing-stick' Which Drifted from Alaska to Greenland." AA, (1890), Old Series, 3:233-36.

903 PALMER, J.B. "The Maori Kotaha." JPS, (1957), 66:175-91.

904 PHILLIPPS, W. "A Maori Spear." *Man*, (1952), 52:96.

905 ROTH, H.L. "Spears and Other Articles from the Solomon Islands." *Internationales Archiv für Ethnographie*, (1898), 11:154-61.

906 UHLE, M. "Peruvian Throwing-sticks." AA, (1909), 11:624-27.

907 VIGNATI, M. "The Use of the Spearthrower in Northwestern Argentina." *Archives Ethnos*, (1948), 1 (1).

908 WALKER, E.H. "An Eskimo Harpoon-thrower." *Masterkey*, (1946), 20:193-94.

909 WILD, R.P. "Bone Dagger and Sheath from Obuasi, Ashanti." *Man*, (1935), 35:10-11.

910 WILLIAMS, G.C. "Suggested Origin of the Malay Keris (Dagger) of the Superstitions Attaching to it." *Royal Asiatic Society, Malayan Branch, Journal*, (1938), 15:127-41.

Part II

Geographical

(NOTE: Page numbers following general works refer to warfare.)

Q / NORTH AMERICA

911 ELLIS, E.S. *The Indian Wars of the U.S.* New York: Cassell Publishing Co., 1892.

912 EWERS, J.C. "Primitive American Commandos." *Masterkey*, (1943), 17:117-25.

913 HAMILTON, C. *Sul sentiero di guerra. Scritti e testimonianze degli indiani d'America.* (American Indian Warfare) Milano: Feltrinelli, 1960.

914 SMITH, Marian W. "American Indian Warfare." *Transactions of the New York Academy of Sciences*, (1951), 2nd Series, 13:348-65.

Q.1 / Alaska & Canada

915 ABERCOMBIE, W.R. "A Military Reconnaissance of the Copper River Valley." In *Compilation of Narratives of Explorations in Alaska*, pp. 563-91. Washington, 1877.

916 BOAS, Franz. "Third Report on the Indians of British Columbia." *British Association for the Advancement of Science, Report of the Annual Meeting*, (1891), 421-23.

917 De LAGUNA, Frederica. *The Story of a Tlingit Community*. Washington: *Smithsonian Institution*, (1960), 10:150-58.

918 DENSMORE, Frances. *Chippewa Customs*. Washington: Government Printing Office, 1929, pp. 132-34, 168-69.

919 ——. *Chippewa Music*. v. 2. Washington: Government Printing Office, 1913, pp. 40-42, 60-132, 185-95.

920 DRUCKER, Philip. *The Northern and Central Nootkan Tribes.* Washington: Government Printing Office, 1951, pp. 334-63.

921 EWERS, John C. "Blackfoot Raiding for Horses and Scalps." In Bohannan, *Law and Warfare*, 1967, pp. 327-49.

922 ——. *The Blackfeet: Raiders on the Northwestern Plains.* Norman: University of Oklahoma Press, 1958.

923 ——. *The Horse in Blackfoot Indian Culture.* Washington: Bureau of American Ethnology, Bulletin 159, 1955, pp. 171-214.

924 GRABURN, Nelson H.H. "Eskimo 'Law' in the Light of Self- and Group-interest." Paper presented at the 67th annual meeting of the American Anthropological Association, Seattle, 1968.

925 HALLOWELL, A.I. "Aggression in Saulteaux Society." *Psychiatry,* (1940), 3:395-407.

926 HICKERSON, H. *The Southwestern Chippewa.* American Anthropological Memoir, No. 92, 1962. See chapt. 2, "Warfare and its Ecological Basis."

927 HONIGMANN, John J. *The Kaska Indians: An Ethnographic Reconstruction.* New Haven: Yale University Press, 1954, pp. 93-95.

928 INNOKENTII, Metropolitan of Moscow. *Zapiski ob ostrovakh Unalashkinskago.* Sanktpeterburg, Izdano Izhdiveniem Rossiisko-Amerikanskoi Kompanii. v. 2, 1840, pp. 93-104, 185-86, 279-90, 320-21.

929 JOHNSTON, W. "Chief Kwah's Revenge." *The Beaver,* (1943), 2:22-23.

930 KENNEDY, G.A. "The Last Battle." *Alberta Folklore Quarterly,* (1945), 1:57-60.

931 KNAPP, Frances, and R.L. Childe. *The Thlinkets of Southeastern Alaska.* Chicago: Stone and Kimball, 1896, pp. 35-39.

932 LANE, Kenneth S. "The Montagnais Indians, 1600-1640." *Kroeber Anthropological Society Papers,* (1952), 21-24, 40, 59.

933 Le CLERCQ, Chrestien. *New Relations of Gaspesia.* Toronto: Champlain Society, 1910, pp. 267-70.

934 LEWIS, Oscar. *The Effects of White Contact upon Blackfoot Culture.* Washington, University of Washington Press, American Ethnological Society Monograph No. 6, (1942), "Changes in Warfare," pp. 46-60.

935 LOWIE, R.H. "The Military Societies of the Plains Cree." ICA, (1955), XXXI, i:3-9.

936 McCLINTOCK, W. "Blackfoot Warrior Societies." *Masterkey*, (1937-8), 11:148-58; 12:11-23.

937 McILWRAITH, Thomas F. *The Bella Coola Indians.* v. 2. Toronto: University of Toronto Press, 1948, pp. 338-76.

938 RODNICK, David. "An Assiniboine Horse-raiding Expedition." AA, (1939), 41:611-16.

939 SAPIR, Edward, and M. Swadesh. *Native Accounts of Nootka Ethnography.* Bloomington: Indiana University Research Center in Anthropology, Folklore, and Linguistics, 1955, pp. 413-33.

940 SCHWATKA, F. *Report of Military Reconnaissance made in Alaska in 1883.* Washington, 1900, pp. 96-103.

941 SKINNER, Alanson. *Political and Ceremonial Organization of the Plains Ojibwa.* New York: American Museum of Natural History, 1913, pp. 483-93, 501-2.

942 SLOBODIN, R. "Eastern Kutchin Warfare." *Anthropologica*, (1960), 2:76-94.

943 SWAN, James G. *The Indians of Cape Flattery, at the Entrance to the Strait of Fuca, Washington Territory.* Washington, Smithsonian Institution, 1870, pp. 50-1.

944 WALLIS, Wilson D. and Ruth S. *The Micmac Indians of Eastern Canada.* Minneapolis: University of Minnesota Press, 1955, pp. 212-18.

Q.2 / Eastern & Central United States

945 ADAMS, F.P. "Pipe of Peace." *Narragansett Dawn*, (1935), I:157.

946 ARMSTRONG, P.A. *The Sauks and the Black Hawk War.* Springfield, Illinois: H.W. Rokker, 1887.

947 BRANNON, P.A. "Creek Indian War, 1836-37." AHQ, (1951), XIII: 956-58.

948 CONNELLEY, W.E. "The Sword and Belt at Orion." *Annual Archaeological Report* (see appendix to the report of the Minister of Education, Ontario [Toronto]), (1905), 68-70.

949 EGGLESTON, G.C. *Red Eagle and the Wars with the Creek Indians of Alabama.* New York: Dodd, Mead & Co., 1878.

950 EMERSON, J.N. "The Old Indian Fort Site." *Ontario History* 50, (1958).

951 GEARING, Fred. *Priests and Warriors: Social Structure for Cherokee Politics in the 18th Century.* American Anthropological Association Memoirs, No. 93, (1962).

952 ———. "The Structural Poses of 18th Century Cherokee Villages." AA, (1958), 60:1148-57.

953 GIFFORD, J.C. *Billy Bowlegs and the Seminole War.* Coconut Grove, Florida: The Triangle Co., 1925.

954 HADLOCK, Wendell S. "War Among the Northeastern Woodland Indians." AA, (1947), 49:204.

955 HALBERT, H.S., and T.H. Ball. *The Creek War of 1813 & 1814.* Chicago: Donohue & Henneberry, 1895.

956 HALL, A.H. "The Red Stick War." CO, (1934), XIII:264-93.

957 HEARD, I.V.D. *History of the Sioux War and Massacres of 1862 & 1863.* New York: Harper & Brothers, 1863.

958 HECKEWELDER, John G.E. *An Account of the History, Manners, and Customs of the Indian Nations, Who Once Inhabited Pennsylvania and the Neighboring States.* Philadelphia: Abraham Small for the American Philosophical Society, 1819, pp. 166-75, 208-9.

959 HOEBEL, E. Adamson. "The Comanche Sun Dance and Messianic Outbreak of 1873." AA, (1941), 43:301-3.

960 ———. "Comanches Had Dictators." *El Palacio,* (1940), 17:211-12.

961 ———. *The Political Organization and Law-ways of the Comanche Indians.* Menasha, Wisconsin: American Anthropological Association, 1940, pp. 21-35, 104-5.

962 HUNT, George T. *The Wars of the Iroquois: A Study in Intertribal Trade Relations.* Madison: University of Wisconsin Press, 1940.

963 JAMES, Edwin, ed. *A Narrative of the Captivity and Adventures of John Tanner during Thirty Years Residence among the Chippewa, Ottawa and Ojibwa Tribes.* Minneapolis: University of Minnesota Press, 1956.

964 KERANS, P. "Murder and Atonement in Huronia." *Martyrs' Shrine Message,* (1953), 17(2):46-47, 53-54.

965 KNOWLES, Nathaniel. "The Torture of Captives by the Indians of Eastern North America." PAPS, (1940), 82:151-225.

966 LEE, Nelson. *Three Years Among the Comanches.* Norman: University of Oklahoma Press, 1957, pp. 95, 135-43.

967 LEJEUNE, Paul. "Hiroquois Cruelty." P.H., (1953), 497-98.

968 LINDESTROM, Peter M. *Geographia Americae with an Account of the Delaware Indians Based on Surveys and Notes Made in 1654-1656.* Philadelphia: Swedish Colonial Society, 1925, chapt. 14, "Warfare."

969 LINTON, Ralph. "The Comanche Sun Dance." AA, (1935), 37:420-28.

970 MARYE, W.B. "Warriors' Paths." PA, (1943-4), XIII:4-26, XIV:4-22.

971 MASON, J. "A Brief History of the Pequot War." MHSC, (1819), ser. 2, VIII:120-53.

972 McKERN, Waldo C. *A Winnebago War-bundle Ceremony.* Milwaukee: Public Museum, Year Book No. 8, 1929.

973 MORGAN, Lewis H. *League of the Ho-de-no-sau-nee or Iroquois.* v. 1. New York: Dodd and Mead, 1901, pp. 72, 257, 328-31.

974 ——. "The Iroquois Faith in Treaties." PH, (1953), 530-31.

975 NAROLL, Raoul. "The Causes of the Fourth Iroquois War." *Ethnohistory,* (1969), 16:51-81.

976 NEWCOMB, William W. *The Culture and Acculturation of the Delaware Indians.* Ann Arbor: University of Michigan Press, 1956, pp. 54-56.

977 ORR, C. *History of the Pequot War.* Cleveland: The Helman-Taylor Co., 1897.

978 OTTERBEIN, K.F. "Why the Iroquois Won: An Analysis of Iroquois Military Tactics." *Ethnohistory,* (1964), 2:56-63.

979 OWEN, M.B. "Indian Wars in Alabama." AHQ, (1951), XIII:92-131.

980 PECKHAM, Howard H. *Pontiac and the Indian Uprising.* Princeton, N.J.: University Press, 1947.

981 POND, C.G. "Indian Warfare in Minnesota," CMHS, (1880), III:129-38.

982 RADIN, P. "A Semi-historical Account of the War of the Winnebago and the Foxes." PSHSW, (1914), 191-207.

983 RADIN, Paul, ed. *Crashing Thunder: The Autobiography of an American Indian.* New York: D. Appleton, 1926, pp. 66-68, 79-86.

984 ——. *The Winnebago Tribe.* Washington, U.S. Bureau of American Ethnology, Annual Report No. 37, 1915-1916, pp. 157-62, 255-79, 301-7.

985 ROBERTS, A.H. "The Dade Massacre." FHSQ, (1927), V:123-128.

986 RODDIS, L.H. *The Indian Wars of Minnesota*. Cedar Rapids, Iowa: Torch Press, 1956.

987 SCHEELE, R. *Warfare of the Iroquois and their Northern Neighbors*. Columbia University, Ph.D. thesis, 1950.

988 SHELDON, George. "The Tradition of an Indian Attack on Hadley, Mass., in 1675." *New England Historical and Genealogical Register*, (October 1874).

989 SIPE, C.H. *The Indian Wars of Pennsylvania*. Harrisburg, 1929.

990 SKINNER, A.B. "An Early Long Island Tragedy." *Indian Notes*, (1925), 2:290-91.

991 ———. "War Customs of the Menomini Indians." AA, (1911), 13:229-312.

992 SMITH, W.W., a Lieutenant of the Left Wing. *Sketches of the Seminole War*. Charleston, 1836.

993 SNYDERMAN, G.S. *Behind the Tree of Peace: A Sociological Analysis of Iroquois Warfare*. University of Pennsylvania, Ph.D. thesis, 1948.

994 SULTE, B. "The War of the Iroquois." AAR, (1899), 124-51.

995 SWANTON, John Reed. "Social Organization and Social Usages of the Indians of the Creek Confederacy." Washington: *U.S. Bureau of American Ethnology, Annual Report No. 42* (1924/1925), 339-45, 406-41.

996 SYLVESTER, H.M. *Indian Wars of New England*. 3 vols. Boston: W.B. Clark Co., 1910.

997 TRELEASE, Allen W. *Indian Affairs in Colonial New York: The 17th Century*. Ithaca: Cornell University Press, 1960.

998 TRUMBULL, B. *A Compendiuum of the Indian Wars in New England*. ed. F.B. Hartranft. Hartford: C.A. Goodwin, 1926.

999 WALLACE, Ernest, and E.A. Hoebel. *The Comanches: Lords of the South Plains*. Norman: University of Oklahoma Press, 1952, pp. 48-49, 223, 251-66.

1000 WAMDITANKA, Big Eagle. "A Sioux Story of the War." CMHS, (1894), VI:382-400.

1001 WHITTELSEY, C. "Indian Affairs around Detroit in 1706." In W.W. Beach, ed., *Indian Miscellany*. Albany: Munsell, 1877, pp. 270-79.

1002 WITHERS, A.S. *Chronicles of Border Warfare*. Cincinnati: The R. Clark Co., 1895.

Q.3 / Great Plains

1003   ALTER, J.C. "Black Hawk's Last Raid." UHQ, (1931), IV:99-108.

1004   ARNOLD, R.R. *Indian Wars of Idaho.* Caldwell, 1932.

1005   BOWERS, Alfred W. *Mandan Social and Ceremonial Organization.* Chicago: University of Chicago Press, 1950, pp. 64-65, 170-73, 308-9.

1006   BRIMLOW, G.F. *The Bannock Indian Wars of 1878.* Caldwell, Idaho: The Caxton Printers, 1938.

1007   BRININSTOOL, E.A. *Fighting Red Clouds Warriors.* Columbus: The Hunter-Trader-Trapper Co., 1926.

1008   BYRNE, P.E. *Soldiers of the Plains.* New York, 1926.

1009   DENIG, Edwin Thompson. "Of the Crow Nation." In E.T. Denig, *Five Indian Tribes of the Upper Missouri.* Norman: University of Oklahoma Press, 1961, pp. 144-48, 162-203.

1010   De SMIT, Rev. P.J. "Indian Warfare." In his *Western Missions and Missionaries.* New York, T.W. Strong, 1859.

1011   DORSEY, G.A. "The Osage Mourning-war Ceremony." AA, (1902), n.s., IV:404-11.

1012   DORSEY, George A. "The Ponca Sun Dance." Field Columbia Museum, *Anthropological Series*, (1905), 7:61-88.

1013   DORSEY, James O. "An Account of the War Customs of the Osages, Given by Red Corn (Hapa Oulse), of the Tsiou Peace-making Gens." *American Naturalist*, (1884), 18:113-33.

1014   ———. *Omaha Sociology.* Washington: Bureau of Ethnology, Annual Report No. 3 (1881-1882), 312-27.

1015   EMMITT, R. *The Last War Trail: The Utes and the Settlement of Colorado.* Norman: University of Oklahoma Press, 1972.

1016   EWERS, J.C. "Review of Hassrick, *The Sioux: Life and Customs of a Warrior Society.*" *Ethnology*, (1964), 11:287-88.

1017   FLANNERY, Regina. *The Gros Ventres of Montana: Part 1, Social Life.* Washington, D.C.: Catholic University of America, 1953, pp. 89-100, 183, 191.

1018   FLETCHER, Alice and F. La Flesche. *The Omaha Tribe.* Washington: Bureau of American Ethnology, Annual Report No. 27, 1905-1906, 405-8, 416-46.

1019 GRINNELL, G.B. *The Fighting Cheyennes.* Norman: University of Oklahoma Press, 1956.

1020 ——. "Lone Wolf's Last War Trip." M, (1943), 17:162-67, 219-24.

1021 ——. "Wart..e of the Plains Indians." In Kroeber, A.L. and T.T. Waverman, *Source Book in Anthropology.* Berkeley, 1920, pp. 421-27.

1022 HALEY, J.E. "The Great Comanche War Trail." PPHR, (1950), XXIII:11-21.

1023 HARVARD UNIVERSITY, Department of Anthropology. *Warfare among the Omaha.* Cambridge: Social Anthropology of Communities, No. 14, mimeographed, 1937.

1024 HASSRICK, Royal B. *The Sioux: Life and Customs of a Warrior Society.* Norman: University of Oklahoma Press, 1964.

1025 JOHNSON, W.F. *Life of Sitting Bull and History of the Indian War.* Philadelphia: Edgewood Publishing Co., 1891.

1026 KROEBER, Alfred L. *Ethnology of the Gros Ventre.* New York: American Museum of Natural History, 1908, pp. 191-219.

1027 La FLESCHE, F. *War Ceremony and Peace Ceremony of the Osage Indians.* Washington, D.C., Bulletins of the Bureau of American Ethnology, no. 101.

1028 LANDES, R. "Dakota Warfare." SWJA, (1959), 15:43-52.

1029 LAYMON, O.F. *Tribal law for the Oglala Sioux.* Pierre, 1953.

1030 LOWIE, Robert Harry. *The Crow Indians.* New York: Farrar and Rinehart, 1935, pp. 9-12, 196-98, 220-21, 230-36, 270-76.

1031 ——. *The Religion of the Crow Indians.* New York: American Museum of Natural History, 1922, pp. 361-68, 383, 394-424.

1032 ——. *The Sun Dance of the Crow.* New York: American Museum of Natural History, 1915.

1033 ——. *Military Societies of the Crow Indians.* New York: American Museum of Natural History, 1913, pp. 158-62, 175, 180-89.

1034 MacGREGOR, G. *Warriors without Weapons.* Chicago: University of Chicago Press, 1946.

1035 MARQUIS, T.B. *A Warrior Who Fought Custer.* Minneapolis: The Midwest Co., 1931.

1036 McALLESTER, David P. "Water as a Disciplinary Agent among the Crow and Blackfoot." AA, (1941), 43:593-604.

1037 MISHKIN, B. *Rank and Warfare among the Plains Indians.* Seattle: University of Washington Press, 1966.

1038 MURIE, J.R. "A Pawnie War Party." PH, (1953), 504-7.

1039 ——. *Pawnee Indian Societies.* New York: American Museum of Natural History, 1914, pp. 560, 573, 580, 595-97, 640.

1040 NEWCOMB, W.W. "A Re-examination of the Causes of Plains Warfare." AA, (1950), 52:317-30.

1041 OPLER, M.K. "The Ute Indian War of 1879." EP, (1939), XLVI:255-62.

1042 RICHARDSON, Rupert N. *The Comanche Barrier to South Plains Settlement; a Century and a Half of Savage Resistance to the Advancing White Frontier.* New York: Kraus, original 1933.

1043 SECOY, F.R. *Changing Military Patterns of the Great Plains.* American Ethnological Society, Monograph No. 21, 1953.

1044 SKINNER, A.B. "An Osage War Party." *Milwaukee, Public Museum Yearbook, 1922,* (1923), 2:165-69.

1045 SMITH, M.W. "The War Complex of the Plains Indians." *Proceedings of the American Philosophical Society,* (1938), 78:425-64.

1046 SPRAGUE, Marshall. *Massacre: The Tragedy at White River.* Boston: Little, Brown and Co., 1957.

1047 TIXIER, Victor. *Tixier's Travels on the Osage Prairies.* Norman: University of Oklahoma Press, 1940, pp. 177-82, 217-28, 260-61.

1048 TROBRIAND, P.R. *Military Life in Dakota.* St. Paul: Alvord Memorial Commission, 1951.

1049 VIGNESS, D.M. "Indian Raids on the Lower Rio Grande, 1836-1837." SWHQ, (1955), LIX:14-23.

1050 VOGET, F.W. "Warfare and the Integration of Crow Indian Culture." In W.H. Goodenough, ed., *Explorations in Cultural Anthropology.* New York: McGraw-Hill, 1964.

1051 WISSLER, Clark. "The Influence of the Horse in the Development of Plains Culture." AA, (1914), 16:1-25.

Q.4 / North & Southwest

1052 ANONYMOUS. "The Modoc Massacre." *Harpers Weekly* (April 26, 1873), 17 (852).

1053 BARTLETT, K. "Hopi History, II: The Navajo Wars." *Memoirs of the American Museum of Natural History*, (1937), 8:33-37.

1054 BASAURI, C. "La resistencia de los Tarahumaras." *Mexican Folkways*, (1926), 2:40-44.

1055 BASSO, Keith H. *Western Apache Raiding and Warfare: From the Notes of Grenville Goodwin.* Tucson: University of Arizona Press, 1971.

1056 BEAGLEHOLE, E. *Notes on Hopi Warfare.* American Anthropological Association, Memoir No. 44, (1935).

1057 BLEDSOE, A.J. *The Indian Wars of the Northwest.* San Francisco: Bacon & Company, 1885.

1058 BLOOM, L.B. "A Campaign against the Moqui Pueblos." NMHR, (1931), 6:158-226.

1059 BRADY, C.T. *Northwestern Fights and Fighters.* New York: The McClure Co., 1907.

1060 CLINE, Walter B., R. Commons, M. Mandelbaum, R. Post, and L.V. Walters. *The Sinkaietk or Southern Okanagon of Washington.* Edited by Leslie Spier. Menasha: George Banta, 1938, pp. 55, 78, 80-83, 92-93, 116, 160-61.

1061 CODERE, H. "Fighting with Property." MAES, (1950), XVIII.

1062 COLLINSON, W.H. *In the Wake of the War Canoe.* London, 1915.

1063 CONQUEST, F.L. "The Cultural Ecology of Economic Systems: A Formal Account of War and Peace in the Southwest." SWAA-AES, (1971).

1064 DEANS, J. "A Strange Way of Preserving Peace amongst Neighbors." AAOJ, (1888), X:42-43.

1065 DILLON, Richard H. "Costs of the Modoc War." *California Historical Society Quarterly*, (1949), 28 (2).

1066 DOUGLAS, G. "Revenge at Guayasdums." *Beaver* (Winnipeg), (September 1952), 283:6-9.

1067 DUNN, J.P. *Massacres in the Mountains: A History of the Indian Wars of the Far West.* New York: Harper Brothers, 1886.

1068 ELIAS, E.E. "El Terrible Veneno: Tactica guerra de los Indios Apaches." *Sociedad Chihuahuense de Estudios Historicos, Boletin,* (1950), 7:392.

1069 ELLIS, F.H. "Patterns of Aggression and the War Cult in Southwestern Pueblos." SWJA, (1951), 7:177-201.

1070 FARMER, M.F. "A Suggested Typology for Defensive Systems of the Southwest." SWJA, (1957), 13:249-267.

1071 FATHAUER, G.H. "The Structure and Causation of Mohave Warfare." SWJA, (1954), 10:97-118.

1072 FITZGERALD, Maurice. "The Modoc War: Reminiscences of the Campaign against the Indian Uprising in the Lava Beds at Northern California and Southern Oregon in 1872-73." *Americana,* (1927), 21 (4).

1073 FONTANA, B.L. "Lost Arsenal of the Papagos." *Desert Magazine,* (1960), 23:22-23.

1074 FREED, Stanley. *Changing Washo Kinship.* Berkeley: University of California Press, 1960, pp. 350-53.

1075 GAYTON, Anna Hadwick. *Yokuts and Western Mono Ethnography.* Berkeley: University of California Press, 1948, pp. 9-11, 145, 154, 160, 176-77, 215.

1076 GIFFORD, Edward W. *Culture Element Distributions: v. 4, Pomo.* Berkeley: University of California Press, 1937, pp. 155, 198, 252.

1077 GOLDSCHMIDT, Walter; George Foster; and Frank Essene. "War Stories from Two Enemy Tribes." JAF, (1939), 52:141-54.

1078 GRINNELL, G.B. "Lone Wolf's Last War Trip." M, (1943), 17:162-67, 219-24.

1079 HAILE, Berard. *Origin Legend of the Navaho Enemy Way.* New Haven: Yale University Press, 1938.

1080 HARRISON, C. *Ancient Warriors of the North Pacific.* London: H.F. & G. Witherby, 1925.

1081 HILL, Willard Williams. "The Navaho Indians and the Ghost Dance of 1890." AA, (1944), 46:523-27.

1082 ———. *Navaho Warfare.* New Haven: Yale University Press, 1936.

1083 HODGE, F.W. "War God Idols at San Juan." *El Palacio,* (1927), 23:588-89.

1084 HOWARD, Helen A. *War Chief Joseph.* Caldwell, Idaho: Caxton Printers, 1941.

1085 HUMPHREY, N.B. "The Mock Battle Greeting." JAFL (1941), 54:186-90.

1086 ILIFF, Flora Gregg. *People of the Blue Water: My Adventures among the Walapai and Havasupai Indians.* New York: Harper, 1954, pp. 162-65.

1087 JONES, L.F. *Indian Vengeance.* Boston: The Stratford Co., 1920.

1088 KROEBER, A.L. "A Kato War." In *Festschrift P.W. Schmidt,* (1928). Also in R.F. Heizer and M. Whipple, *The California Indians,* (1960), pp. 397-403.

1089 ———. "Law of the Yurok Indians." ICA, (1926), XXII, ii:511-16.

1090 ———. "A Yurok War Reminiscence." SWJA, (1945), I:318-32.

1091 ———. *Handbook of the Indians of California.* Washington: Government Printing Office, 1925, pp. 49-52.

1092 KURTZ, Ronald J. "Headman and War Chanters: Role Theory and the Early Canyoncita Navajo." *Ethnohistory,* (1969), 16:83-111.

1093 LEIGHTON, Dorothea, and J. Adair. *People of the Middle Place: A Study of the Zuni Indians.* New Haven, Human Relations Area Files, Unpublished Manuscript, 1963, pp. 32, 75, 82.

1094 LINTON, R. "Nomad Raids and Fortified Pueblos." *American Antiquity,* (1944), 10:28-33.

1095 LOEB, Edwin Meyer. *Pomo Folkways.* Berkeley: University of California Press, 1926, pp. 200-11.

1096 MEACHAM, Alfred. *Wigwam and War-path, or the Royal Chief in Chains.* Boston: John P. Dale and Co., 1875.

1097 MURRAY, Keith A. *The Modocs and Their War.* Norman: The University of Oklahoma Press, 1959.

1098 NEQUATEWA, E.C. "A Mexican Raid on the Hopi Pueblo of Oraibi." *Plateau,* (1944), 16:44-52.

1099 OBERG, K. "Crime and Punishment in Tlingit Society." AA, (1943), n.s., XXXVI:145-56.

1100 OPLER, M.E. "A Jicarilla Apache Expedition and Scalp Dance." JAFL (1941), 54:10-23.

1101 ———. *Dirty Boy: A Jicarilla Tale of Raid and War.* American Anthropological Association Memoir No. 52, 1938.

1102 OPLER, Morris Edward, and Harry Hoijer. "The Raid and Warpath Language of the Chiricahua Apache." AA, (1940), 42:617-34.

1103 PARSONS, Elsie C. *The Social Organization of the Tewa of New Mexico.* Menasha: American Anthropological Association, 1929, pp. 137-38, 242.

1104 PRICE, John A. *Washo Economy.* Carson City: Nevada State Museum, 1962, pp. 7, 30, 51, 56.

1105 RAY, Verne F. *Cultural Relations in the Plateau of North America.* Los Angeles: Southwest Museum, 1939, pp. 36-45.

1106 RIDDLE, J.C. *The Indian History of the Modoc War.* San Francisco: Marnell & Co., 1914.

1107 RILEY, C.L. "Defensive Structures in the Hovenweep Monument." *El Palacio,* (1950), 57:339-44.

1108 SANTEE, J.F. "Edward R.S. Canby, Modoc War, 1873." *Oregon Historical Quarterly,* (1932),33 (1).

1109 SCHMITT, M.F., and D. Brown. *Fighting Indians of the West.* New York: C. Scribner, 1948, pp. 229-50.

1110 SMITH, Marian W. "Review of H. Howard, *War Chief Joseph.*" AA, (1942), 44:302.

1111 SMITH, Watson and J.M. Roberts. *Zuni Law: A Field of Values.* Cambridge: Harvard University Press, 1954, pp. 156-58.

1112 SPIER, Leslie. "Maricopa Warfare." In R.C. Owen, et al. *The North American Indians,* New York, 1967, pp. 453-59.

1113 ———. *Yuman Tribes of the Gila River.* Chicago: University of Chicago Press, 1933, pp. 132, 135, 154, 162, 166-79, 292-93.

1114 ———. *Havasupai Ethnography.* New York: American Museum of Natural History, 1928, pp. 178-79, 248-56, 256-80.

1115 ———. "The Ghost Dance of 1870 among the Klamath of Oregon." *University of Washington Publications in Anthropology,* (1928), 2:39-56.

1116 STEWARD, K.M. "Mohave Warfare." SWJA, (1947), 3:257-78.

1117 STRATTON, R.B. *Captivity of the Oatman Girls.* New York, 1857.

1118 STUART, B.R. "Paiute Surprise in the Mohave." *Masterkey,* (1943), 17:217-19.

1119 SWADESH, M. "Motivations in Nootka Warfare." SWJA, (1948), IV:76-93.

1120 TEIT, James A. *The Salishan Tribes of the Western Plateau. The Okanagon.* Washington: Bureau of American Ethnology, Annual Report No. 45, 1927/1928, 257-73, 389-90.

1121 THOMPSON, Lucy. *To the American Indian.* Eureka, California: Cummins Print Shop, 1916, pp. 143-44.

1122 TURNER, Christy G., and N.T. Morris. "A Massacre at Hopi." *American Antiquity*, (1970), 35:320-31.

1123 VICTOR, F.F. *The Early Indian Wars of Oregon.* Salem: F.C. Baker, 1894.

1124 VOGT, Evon Z. *Navaho Veterans: A Study of Changing Values.* Cambridge: Harvard Peabody Museum Papers, 1951, v. 41.

1125 WALLACE, W.J. *Hupa Warfare.* Los Angeles: Southwest Museum, 1949, Leaflet No. 23.

1126 WEBER, A. "Navajos on the Warpath." FMSW, (1918), VII, 1-18.

1127 WHEELER-VOEGELIN, Erminie. "The Northern Paiute of Central Oregon: A Chapter in Treaty Making." *Ethnohistory*, (1955), 2:95-132, 241-72; (1956), 3:1-10.

1128 ——. *Tubatulabal Ethnography.* Berkeley: University of California Press, 1938, pp. 49-50.

1129 WHITE, Cris W. "Lower Colorado Tribal Warfare: A Reconsideration." SWAA-AES, (1971).

1130 WOODBURY, Richard B. "A Reconsideration of Pueblo Warfare in the Southwest U.S." *Actas del XXXIV congreso internacional de Americanistas.* San Jose, Costa Rica, (1959), 2:124-133.

1131 WORCHESTER, Donald E. "The Navajo during the Spanish Regime in New Mexico." NMHR, (1951), 25:101-18.

Q.5 / Mexico

1132 BANDELIER, Adolph F. *On the Art of War and Mode of Warfare of the Ancient Mexicans.* Cambridge: Harvard University, Peabody Museum of American Archaeology and Ethnology, Report No. 2, 1876/1879.

1133 BANNON, John. "The Conquest of the Chinipas." *Mid-America*, (1939), 21:3-31.

1134 BORGONIO, Gaspar Guadelupe. "Organizacion militar de los Tenochca." *Revista Mexicana de estudios anthropologica*, Mexico, (1954-55), 14:381-83.

1135 CANESCO, Vincourt. *La Guerra Sagrada.* Mexico: Instituto Nacional de Anthropologia e Historia, 1966.

1136 CRUMRINE, Lynne S., and N. Ross Crumrine. "Blood Sacrifice Mediating Binary Oppositions: A Structural Analysis of Two Mayo Myths." Paper presented at the 67th annual meeting of the American Anthropological Association, Seattle, 1968.

1137 DAVIS, Edward H., and E.Y. Dawson. "The Savage Seris of Sonora." *Scientific Monthly*, (1945), 60:193-202, 261-68.

1138 DENSMORE, Frances. *Papago Music.* Washington: Government Printing Office, 1929, pp. 112, 175-86, 192.

1139 DUNNE, Peter. "The Tepehuan Revolt." *Mid-America*, (1936), 18:3-14.

1140 KATZ, F. "The Causes of War in Aztec Mexico." *Wiener Völkerkundliche Mitteilungen.* (Wien, 1955), 3:31-33.

1141 KELLY, W.H. "The Place of Scalps in Cocopa Warfare." *El Palacio*, (1949), 56:85-91.

1142 KELLY, William H. *The Papago Indians of Arizona: A Population and Economic Study.* Tucson: Bureau of Ethnic Research, Department of Anthropology, University of Arizona, 1963, pp. 41-45.

1143 LEWIS, Oscar. *Life in a Mexican Village: Tepoztlan Restudied.* Urbana: University of Illinois Press, 1951, pp. 4-5, 47-49, 56, 77, 117-18, 127, 211, 268, 272, 297, 429.

1144 McGEE, W.J. *The Seri Indians.* Washington: Bureau of American Ethnology, Annual Report No. 17, 1895-1896, pp. 154, 259-65.

1145 MEANS, Philip A. *History of the Spanish Conquest of Yucatan and of the Itzas.* Cambridge: Harvard Peabody Museum Papers, 1917, v. 7.

1146 MONOZ CAMARGO, D. "Los flaxcaltecas, arte militar, arras ofensivas y defensivas, prisioneros de guerra, sacrificios humanos." *El Impulsor Bibliografico*, (1943), 9:1-3.

1147 MORIARTY, James R. "The Origin and Development of Maya Militarism." *Colorado State College, Museum of Anthropology, Miscellaneous Series* No. 9:23-33.

1148 MOSER, Edward. "Seri Bands." *Kiva*, (1963), 3:16, 20, 22-27.

1149 NASH, June Caprice. "Death as a Way of Life: The Increasing Resort to Homicide in a Maya Indian Community." AA, (1967), 69:455-70.

1150 ORELLANA, T. Rafael. "La guerra." *Esplendor del Mexico Antiquo*, (1959), 2 vols. (Noriega, et al., eds.) Mexico.

1151 REDFIELD, Robert. *A Village that Chose Progress: Chan Kom Revisited.* Chicago: Univeristy of Chicago Press, 1962, pp. 2-20, 129-36.

1152 ———. *Tepoztlan, A Mexican Village*. Chicago: University of Chicago Press, 1930, pp. 55-65, 99, 103.

1153 ———, and A. Villa Rojas. *Chan Kom: A Maya Village*. Chicago: University of Chicago Press, 1934, pp. 8-11, 20-29, 65, 219-44, 281-87, 318-23.

1154 SAHAGUN, Bernardino de. *Florentine Codex. Book 12—The Conquest of Mexico*. Sante Fe: School of American Research and University of Utah, 1955, pp. 35, 39, 44n-v, 53-115.

1155 ———. *Florentine Codex: General History of the Things of New Spain, Book 8—Kings and Lords*. Santa Fe: School of American Research and University of Utah, 1954, pp. 36, 52-56, 65, 88.

1156 SPICER, Edward H. "Yaqui Militarism." *Arizona Quarterly*, (1947), 3:40-48.

1157 THOMPSON, J. Eric. *Ethnology of the Mayas of Southern and Central Honduras*. Chicago: Field Museum of Natural History, 1930, pp. 38-39.

1158 UNDERHILL, Ruth M. *Papago Indian Religion*. New York: Columbia University Press, 1946, pp. 166, 171-93.

1159 ———. *Social Organization of the Papago Indians*. New York: Columbia University Press, 1939, pp. 132-36.

1160 VALLIANT, George C. *Aztecs of Mexico: Origin, Rise and Fall of the Aztec Nation*. Garden City, New York: Doubleday, 1941, pp. 216-23, 240, 244, 250, 255.

1161 VILLA ROJAS, Alfonso. *The Maya of East Central Quintana Roo*. Washington: Carnegie Institution of Washington, 1945, pp. 44-45, 63, 67-68, 77, 106-7, 113-15.

R / SOUTH AMERICA

1162 ANONYMOUS. "Chamacocos on the Warpath." IS, (1909), IV, 62.

1163 ARMSTRONG, J.M., and A. Metraux. "The Guajira." HSAI, (1948), 4:369-83.

1164 ARNOTT, John. "Los Toba-Pilaga del Chaco y sus guerras." Buenos Aires, *Revista Geografic Americana*, (1934), 1:7:491-501.

1165 BALDUS, Herbert. "Tribos da bacia do Araguaia e o servico de protecao dos Indios." Sao Paulo: *Museu Paulista, Revista*, (1948), 2:137-68.

1166 ———. "Os Tapirape: Tribu Tupi no Brasil Central." Sao Paulo: *Arquivo Municipal, Revista*, (1944-1949), pp. 96-127.

1167 BARKER, James. "Las Incursiones Entre los Guaika." *Boletin Indigenista Venezolano*, (1959), 7:151-67.

1168 ——. "Memoria sobre la cultura de los Guaika." *Boletin Indigenista Venezolano*, (1953), 1:433-89.

1169 BARROS, A. *La guerra contra los Indios.* Buenos Aires, 1875.

1170 BERNAL VILLA, Segundo. "Economia de los Paez." *Revista Colombiana de Antropologia*, (1954), 3:291-367.

1171 BIOCCA, Ettore. *Yanoama: The Narrative of a White Girl Kidnapped by Amazonian Indians.* New York: Dutton, 1970.

1172 BOURNE, Benjamin F. *The Captive in Patagonia; or, Life among the Jiants.* Boston: D. Lathrop, 1874, pp. 60, 142, 149-50.

1173 BRAM, J. *Analysis of Inca Militarism.* American Ethnological Society, No. 4, 1941.

1174 BRETON, Raymond, and Armand de la Paix. *Relation de l'Ille de la Guadeloupe.* In Joseph Rennard, ed., *Les Caraibes, La Guadeloupe, 1635-1656.* Paris: Libraire Générale et Internationale, 1929, pp. 19-23, 45-74.

1175 CARRERA ANDRADE, J. "La guerra de los Incas contra los 'hombres de los Nubes.' " *Cuadernos Americanos*, (1959), CII, i, 145-60.

1176 CHAGNON, N.A. "The Culture-Ecology of Shifting (Pioneering) Cultivation among the Yanomamo Indians." *Proceedings of the 8th International Congress of Anthropological and Ethnological Sciences*, (1970), 3:249-55.

1177 ——. "The Feast." NH, (1968), 77:4, 34-41.

1178 ——. "Personal Fierceness and Headmen in Yanomamo Disputes." Paper presented at the 67th annual meeting of the American Anthropological Association, Seattle, 1968.

1179 ——. *Yanomamo: The Fierce People.* New York: Holt, Rinehart and Winston, 1968.

1180 ——. "Yanomamo Social Organization and Warfare." In M. Fried, *War*, (1968), 109-59.

1181 ——. "Yanomamo—the Fierce People." NH, (1967), 76:22-31.

1182 ——. "Yanomamo Warfare, Social Organization and Marriage Alliances." University of Michigan, Ph.D. thesis, 1966.

1183 CIEZA de LEON, P. de. "The War of Chupas." HSS, (1918), XLII.

1184 ——. "The War of Las Salinas." HSS, (1923), LIV.

1185 ———. "The War of Quito." HSS, (1913), XXXI.

1186 ———. *The Travels of Pedro de Cieza de Leon, A.D. 1532-50.* Trans. by C.R. Markham, 2 v. London: Hakluyt Society, 1869-1871, pp. 653-58, 660-68.

1187 ———. *The Travels of Pedro de Cieza de Leon, A.D. 1532-50, Contained in the First Part of His Chronicle of Peru.* Trans. by C.R. Markham. London: Hakluyt Society, 1864, 2 v., pp. 164-65, 420-21, 424, 501-3, 507-8, 512-13, 516-18, 548-50, 559-65, 629-30, 641, 644-49, 688-700, 710-16.

1188 COOPER, John M. "The Yahgan." HSAI, (1946), 1:81-106.

1189 COTLOW, L. *Amazon Head-hunters.* New York: Holt, 1953.

1190 DICKEY, Herbert Spencer. "The Head Shrinkers of Ecuador." *Masterkey,* (1936), 10:201-3.

1191 DIVALE, William T. "Population Control and the Yanomamo Women Shortage: An Explanation for Amazonian Warfare." Paper presented at the Spring 1970 annual meeting of the Southwestern Anthropological Association, Asilomar, California, 1970.

1192 DOBRITZHOFER, Martin. *An Account of the Abipones, an Equestrian People of Paraguay,* v. 2. Trans. from the Latin by Sara Coleridge. London: J. Murray, 1822, pp. 55, 70, 76, 104-5, 130, 149-50, 350-51, 358-94, 399, 413-22.

1193 DUTERTRE, Jean Baptiste. *Histoire générale des Antilles habitées par les Francais.* v. 2. Paris: T. Jolly, 1667, pp. 2-3, 35-37.

1194 DYOTT, G.M. *Manhunting in the Jungle.* Indianapolis: The Bobbs-Merrill Co., 1930.

1195 FERNANDES, Florestan. *A funcio social de la guerra na sociedade Tupinamba.* Sao Paulo, Brazil, 1952.

1196 ———. "Le Guerre et le sacrifice humain chez les Tupinamba." *Journal de la société des Américanistes de Paris,* (1952), n.s., 41:139-220.

1197 ———. "A analise functionalizta de guerra: Possibilidades de aplicaçao a sociedade Tupinamba." *Museu Paulizta, Revista,* (1949), 3:7-128.

1198 FISHER, H.T. *The Last Inca Revolt.* Norman: University of Oklahoma Press, 1966.

1199 FOCK, Niels. "Mataco Marriage." *Folk,* (1963), 5:91-101.

1200 GILT CONTRERAS, M.A. "Las guerillas indigenas de Chiyarage i Togto." *Archivos Peruanos de Folklore*, (1955), 1:110-19.

1201 GUSINDE, Martin. *Die Yamana: vom Leben und Denken der Wassernomaden am Kap Hoorn.* Modling bei Wien, Anthropos-Bibliothek, 1937, pp. 995-1000, 1137.

1202 GUTIERREZ de PINEDA, Virginia. *Organizacion social en la Guajira.* Bogota, 1950, pp. 1, 191, 201-9.

1203 GUTIERREZ de SANTA CLARA, P. *Historia de las guerras civiles del Peru.* 6 vols. CLD:II, IV, X, XX, XXI. 1904-29.

1204 HAGEN, V.W. von. *Off with Their Heads.* New York: The Macmillan Co., 1937.

1205 HARNER, Michael J. "Jivaro Souls." AA, (1962), 64:258-72.

1206 HENRY, Jules. *Jungle People: A Kaingang Tribe of the Highlands of Brazil.* New York: J.J. Augustin, 1941, pp. 51-60, 89-90, 108-11.

1207 HERSKOVITS, Melville J. and Frances S. *Rebel Destiny: Among the Bush Negroes of Dutch Guiana.* New York: McGraw-Hill, 1934.

1208 HURAULT, Jean. *Les Noirs refugiés Boni de la Guyane française.* Dakar: Institut Français d'Afrique Noire, 1961, pp. 23, 142, 226, 233-34, 293.

1209 KARSTEN, Rafael. "Blood Revenge and War among the Jibaro Indians of Eastern Ecuador." In Bohannan, *Law and Warfare*, (1967), 303-326.

1210 ——. *Indian Tribes of the Argentine and Bolivian Chaco.* Helsingfors: Akademische Buchhandlung, 1932, see chapt. 11, "Warfare."

1211 ——. *The Head-Hunters of Western Amazonas.* Helsingfors, Centraltryckeriet, 1935.

1212 ——. *Blood Revenge, War, and Victory Feasts among the Jibaro Indians of Eastern Ecuador.* Washington: Bureau of American Ethnology, Bulletin No. 79, 1923.

1213 KELM, H. "Die Sitte de Pfeilduells bei den Yuracare (Ostbolivien)." Berlin, *Baessler-Archiv*, (1965), 37:281-310.

1214 LATCHAM, Ricardo E. "La capacidad guerrera de los Araucanos: sus armas y metodos militares." *Revista Chilena de Historia y Geografia*, 15:19:22-93. Santiago: Imprenta universitana, 1915.

1215 ——. "Ethnology of the Araucanos." JAI, (1909), 39:334-70.

1216 LERY, Jean de. *Extracts out of the Historie of John Lerius a Frenchman, Who Lived in Brasill with Mons. Villagagnon, ann. 1557 and 58.* Ed. by Samuel Purchas. *Hakluytus Posthumus or Purchas His Pilgrimes,* (1906), 16:518-79.

1217 LIPKIND, William. "The Caraja." HSAI, (1948), 3:179-91.

1218 MARQUEZ MIRANDA, F. "Los Diaquitas y la guerra." Mendoza: Universidad Nacional de Cuyo, *Instituto de Etnografia Americana, Anales,* (1943), 3:83-117; 4:47-66.

1219 MENDOZA, L.G. "Guerra civil entre Vascongados y otras naciones de Potosi." *Cuadernos de la Coleccion de la Cultura Boliviana,* (1955), V, Potosi.

1220 METRAUX, A. "Tupinamba—War and Cannibalism." HSAI, (1948), 3:119-26.

1221 ——. "The Caingang." HSAI, (1946), 1:445-75.

1222 ——. "Myths and Tales of the Matako Indians (the Gran Chaco, Argentina)." *Ethnological Studies,* (1939), 1-127.

1223 ——. "Études sur la civilisation des Indiens Chiriguano." Tucuman: Universidad Nacional, *Instituto de Ethnologia, Revista,* (1929/1930), 1:295-493 (see pp. 314-16, 319, 327, 364).

1224 MURPHY, R.F. *Headhunter's Heritage: Social and Economic Change among the Mundurucu Indians.* Berkeley, University of California Press, 1960, pp. 127-31, 137-38, 155.

1225 ——. "Reply to H.C. Wilson's Reply on the Causes of Mundurucu Warfare." AA, (1958), 60:1196-99.

1226 ——. "Intergroup Hostility and Social Cohesion." AA, (1957), 59:1018-35.

1227 ——. "Mundurucu Indians, A Dual System of Ethics." In Vergilius Ferm, ed., *Encyclopedia of Morals.* New York: Philosophical Library, 1956, pp. 369-74.

1228 ——. "Aboriginal Culture." In his, *Rubber Trade and the Mundurucu Village.* Columbia University, Ph.D. thesis, 1954.

1229 MUSTERS, George Chaworth. *At Home with the Patagonians.* London: John Murray, 1873, pp. 86, 97, 323-34.

1230 NACHITIGALL, Horst. *Tierradentro, Archaologie und Ethnographie einer kolumbianischen Landschaft.* Zurich: Origo, 1955, pp. 117, 145-46, 171, 286.

1231 NIMUENDAJU, Curt. *The Eastern Timbira.* Trans. and ed. by R.H. Lowie. Berkeley: University of California Press, 1946, pp. 126, 140, 150-52, 159, 161.

1232 NINO, Bernardino de. *Etnografia Chiriguano.* La Paz: Tipografia Comercial de Ismael Argote, 1912, pp. 67, 98, 120, 151, 160, 219, 275-78, 280-81, 298.

1233 OLASCOAGA, M.J., ed. *La conquête de la Pampa.* Buenos Aires, 1881.

1234 PADDEN, R.C. "Culture Change and Military Resistance in Araucanian' Chile, 1550-1730." SWJA, (1957), XIII:103-21.

1235 PALERM, Angel. "Notas sobre las construcciones militares y guerra en Mesoamerica." Mexico, *Anales del Instituto Nacional de Antropologio e Historia,* (1956), 8:37:123-34.

1236 PAYNE, J.L. *Patterns of Conflict in Columbia.* New Haven: Yale University Press, 1968.

1237 PETRULLO, Vincenzo. "Composition of 'Torts' in Guajira Society." *Philadelphia Anthropological Society, Publications,* (1937), 1:153-60.

1238 PRESCOTT, W.H. *History of the Conquest of Peru.* 2 vols. New York: Harper & Brothers, 1847.

1239 REISS, W. "Ein Besuch bei de Jivaros-Indianern." *Gesellschaft für Erdkunde zu Berlin, Verhandlungen,* (1880), 7:325-37.

1240 ROUSE, Irving. "The Carib." HSAI, (1948), 4:547-65.

1241 ROWE, John Howland. "Inca Culture at the Time of the Spanish Conquest." HSAI, (1946), 2:183-330.

1242 SCHORR, Thomas. "The Ecology of Violence as a Way of Life in Rural Columbia." Paper presented at the 69th annual meeting of the American Anthropological Association, San Diego, 1970.

1243 SCHUSTER, Meinhard. "Die Soziologie de Waika." *International Congress of Americanists, Proceedings,* (1956), 32:114-22.

1244 SIMONS, F.A.A. "An Exploration of the Guajira Peninsula, U.S. of Colombia." *Royal Geographical Society, Proceedings,* (1885), 7:781-96, 840.

1245 STADEN, Hans. *The True Story of His Captivity.* Trans. by Malcolm Letts. London: G. Routledge, 1928, pp. 39-41, 62-64, 104-7, 153-54.

1246 STEWARD, Julian H., ed. *Handbook of South American Indians.* v. 7, Index. (See, Feuds, p. 95; Revenge, p. 218, Warfare, p. 276-77.)

Washington: Smithsonian Institution, Bureau of American Ethnology, Bulletin No. 143, vs. 1-7.

1247 STIRLING, Matthew W. *Historical and Ethnographical Material on the Jivaro Indians.* Washington: Government Printing Office, 1938, pp. 39, 42, 51-55, 60-61, 115-17.

1248 STONE, Doris Zemurray. *The Talamancan Tribes of Costa Rica.* Cambridge: Peabody Museum, 1962, pp. 47a, b, 48a, b, 49a, b.

1249 TAPSON, A.J. "Indian Warfare on the Pampa during the Colonial Period." HAHR, (1962), XLII:1-28.

1250 TESILLO, S. de. "Guerras de Chile." CHC, (1864), V, iii.

1251 THERET, André. *Les Singularites de la France Antarctique, autrement nommée Amérique.* Paris: Maisanneure, 1878, pp. 185-92.

1252 UP de GRAFF, Fritz W. *Head Hunters of the Amazon.* New York: Duffield, 1923, pp. 247, 250-53, 260-71.

1253 VISO, J. Del. "La conquista del Desierto." RUNC, (1933), XX (i/ii):3-66.

1254 WAGNER, L.D. "Massacre de Jules Crevaux." JSAP, (1910), n.s., VII:121-2.

1255 WILSON, H. Clyde. "Regarding the Causes of Mundurucu Warfare." AA, (1958), 60:1193-96.

1256 ZEBALLOS, E.S. *La conquista de quince mil leguas.* 2d ed. Buenos Aires, 1878.

1257 ZERRIES, Otto. *Waika (Yanomamo): Die Kulturgeschichtlich Stellung de Waika-Indianer (Yanomamo Indians) des oberen Orinoco im Rahmen der Volkerkunde Sudamerikas.* Munich: Venezuela der Frobenius-Expedition 1954/55 nach Sudost-Venezuela, (1964).

S / AFRICA

1258 BEATTIE, J.H.M. "Checks on the Abuse of Political Power in Some African States: A Preliminary Framework for Analysis." *Sociologus,* (1959), 9:97-115.

1259 BERKELEY, George F.H. *The Campaign of Adowa and the Rise of Menglik.* London, 1902.

1260 BOHANNAN, Paul. *African Homicide and Suicide.* Princeton: Princeton University Press, 1960.

1261 GLUCKMAN, Max. *Custom and Conflict in Africa.* New York: Barnes and Noble, 1964.

1262 LEWIS, D.G. "The Battle at Zwangenduba." *Nada,* (Salisbury, 1956), 33:51-52.

1263 MBOYA, T.J. *Conflict and Nationhood.* London: African Bureau, 1963.

1264 PAULME, D. "Le Guerrier: sa place dans une pensée africaine, hier et aujourd'hui." Paris, *Journal de psychologie normale et pathologique,* (1960), 57:275-90.

1265 SEDOLO, M.D. "Considerations sur la guerre de Ketou-Woto-to." *Notes Africaines,* (1955), 65:24-26.

1266 WALTER, Eugene V. *Terror and Resistance: A Study of Political Violence, with Case Studies of Some Primitive African Communities.* New York: Oxford University Press, 1969.

S.1 / West Africa

1267 AJAYI, J.F.A. "Professional Warriors in Nineteenth Century Yoruba Politics." *Tarikh,* (1965), 1:72-81.

1268 ——, and R. Smith. *Yoruba Warfare in the Nineteenth Century.* Cambridge: Cambridge University Press, 1964.

1269 *Akiga's Story: The Tive Tribe as Seen by One of Its Members.* London: Oxford University Press, 1939, pp. 135-43, 338-42.

1270 ARMITAGE, Cecil Hamilton, and Arthur F. Montanaro. *The Ashanti Campaign of 1900.* London, 1901.

1271 BOCKANI, J. "Mende Warfare." *Farm and Forest,* (1945), 6:104-5.

1272 BOHANNAN, Paul. "The Migration and Expansion of the Tiv." *Africa,* (1954), 24:2-16.

1273 BRACKENBURY, Sir Henry. *The Ashanti War.* Edinburgh and London: W. Blackwood, 1874.

1274 CARMIGNANI, Renzo. *Il cannibalismo degli Asande (o Niam-Naim).* Rome, 1954.

1275 DIVALE, William T. "Ibo Warfare, Social Organization and Population Control." *California Anthropologist,* (California State College at Los Angeles, 1971), 1:10-24.

1276 JEFFREYS, M.D. "The Date of the Bamum-Banso War." *Man,* (1962), 62:70-71.

1277 ———. "Ibo Warfare." *Man,* (1956), 56:77-79.

1278 KOPYTOFF, Igor. "Extension of Conflict as a Method of Conflict Resolution among the Suku of the Congo." JCR, (1961), 5:61-69.

1279 LLOYD, Alan. *The Drums of Kumasi: The Story of the Ashanti Wars.* London, 1964.

1280 OTTERBEIN, K.F. "Higi Armed Combat." SWJA, (1968), 24:195-213.

1281 ———. "Review of Ajayi, 1964, *Yoruba Warfare in the 19th Century.* AA, (1966), 68:1531.

1282 RICKETTS, Major. *Narrative of the Ashantee War.* New York: Barnes and Noble, 1831.

1283 SIDDLE, D.J. "War-towns in Sierra Leone: A Study in Social Change." *Africa,* (1968), 38:1, 47-56.

1284 SKINNER, Elliott. *The Mossi of the Upper Volta.* Stanford: Stanford University Press, 1964, pp. 61-75, 97-106.

1285 TALBOT, P.A. "The Conduct of Oronn Warfare." PH, (1953), 509-19.

S.2 / East & Central Africa

1286 ALLEN, W.E.D. *Guerilla War in Abyssinie.* Harmondsworth, Middlesex, England, and New York: Penguin Books, 1943.

1287 ANDREZEJEWSKI, B.W. "Ideas about Warfare in Borana Galla Stories and Fables." *African Language Studies,* (1962), 3:116-36.

1288 BARBEY, Georges. "Les Massai, patres-guerriers." *Annuel Bulletin,* (Geneva: Musée et Institut d'Ethnographie, 1958), 24-25.

1289 BUGEAU, François. "Les Wakikouyous et la guerre." Rome, *Annali Lateranesi,* (1943), 7:183-226.

1290 CHILDS, Gladwyn M. *Umbundu Kinship and Character.* London: International African Institute and the Witwatersrand University Press, 1949, pp. 194-211.

1291 CZEKANOWSKI, Jan. *Forschungen im Nil-Kongo Zwischengebiet.* Leipzig: Klinkhardt und Biermann, 1924, 2:46-60.

1292 DUNDAS, Charles C. *Kilimanjaro and Its People.* London: Witherby, 1924, pp. 70-100.

1293 ELLIOTT, H.F.I. "The Coiffeur (sic) of the Masai Warrior." TNR, (1948), 26:80-82.

1294 EVANS-PRITCHARD, E.E. "Zande Border Raids." Africa, (1957), 27:217-31.

1295 ———. "Zande Warfare." Anthropos, (1957), 52:239-62.

1296 ———. "A Note on Ghostly Vengeance among the Anuak of the A-E. Sudan." Man, (1953), 53:6-7.

1297 ———. "Ghostly Vengeance among the Luo of Kenya." Man, (1950), 50:87.

1298 ———. "Nuer Curses and Ghostly Vengeance." Africa, (1949), 19:288-92.

1299 ———. The Nuer. Oxford: Clarendon Press, 1940, pp. 142-76.

1300 FALLERS, Lloyd A. Bantu Bureaucracy. A Study of Integration and Conflict in the Political Institutions of an East African People. Cambridge, 1956.

1301 GUTMANN, Bruno. Die Stammeslehren der Dschagga. Munchen: Beck, 1932, pp. 274-83.

1302 ———. Das Recht der Dschagga. Munchen: Beck, 1926, pp. 213-20.

1303 HAMBLY, Wilfrid D. "Warriors of East Africa." Field Museum News, (1938), 8:2.

1304 HANF, T. Die politische Bedeutung ethnischer Gegensätze in Ruanda und Urundi. Freiburg-im-Breisgau: Arnold-Bergstraesser-Institut für Kulturwissenschaftliche Forschung, 1964.

1305 HATCH, John. "Border War." New Statesman, (1964), 67:278.

1306 KAGWA, Apolo. The Customs of the Baganda. New York: Columbia University Press, 1934, pp. 89-91.

1307 KINGDON, F.D. "Bracelet Fighting in the Nuba Mountains." Sudan Notes and Records, (1938), 21:197-200.

1308 KOPYTOFF, Igor. "Extension of Conflict as a Method of Conflict Resolution among the Suku of the Congo." JCR, (1961), 5:61-69.

1309 LINDBLOM, Karl G. "Manière de faire la guerre et usages s'y rattachant chez les Kambas de l'Afrique orientale anglaise." Ymer, (Stockholm, 1914), 123-37.

1310 MACKENZIE, D.R. The Spirit-Ridden Konde. London: Seeley and Service, 1925, pp. 169-77.

1311 MACKENZIE, P.Z. "The Battle of Agrat, 1952." *Sudan Notes and Records*, (1953), 34:309-10.

1312 MEYER, Hans. *Die Barundi: Eine völkerkundliche Studie aus Deutsch-Ostafrika*. Leipzig: Otto Spamer, 1916, pp. 86-88.

1313 OYLER, D.S. "The Shilluk Peace Ceremony." *Sudan Notes and Records*, (1920), 3:296-99.

1314 REDMAYNE, A. "The War Trumpets and Other Mistakes in the History of the Hehe." *Anthropos*, (1970), 65:98-109.

1315 ROSCOE, John. *The Baganda*. London: Macmillan, 1911, pp. 249-59.

1316 WEATHERBY, J.M. "Intertribal Warfare on Mount Elgon in the 19th and 20th Centuries." Kampala, *Uganda Journal*, (1962), 26:200-12.

1317 WILSON, G.M. "Homicide and Suicide among the Joluo of Kenya." In P. Bohannan, ed., *African Homicide and Suicide*, 1960, pp. 179-213, 273-74, 287.

1318 ———. *The Constitution of Ngonde*. Livingstone: Rhodes-Livingstone Institute, 1939, pp. 61-62.

S.3 / South Africa

1319 ANDERSON, Edwin A.H. *With the Mounted Infantry and the Mashonaland Field Force, 1896*. London, 1908.

1320 COLSON, E. "Social Control and Vengeance in Plateau Tonga Society." *Africa*, (1953), 23:199-211.

1321 FURNEAUX, Rupert. *The Zulu War*. Philadelphia: Lippincott, 1963.

1322 GATEHOUSE, R.P. "Military Science Applied to Great Zimbabwe." *South African Journal of Science*, (1963), 59:13-18.

1323 GLUCKMAN, Max. "The Rise of the Zulu Empire." *Scientific American*, (1960), 202:4, 157-168.

1324 ———. "Succession and Civil War among the Bemba: An Exercise in Anthropological Theory." *Rhodes-Livingston Journal*, (1954), 16:6-25.

1325 JUNOD, Henri Alexandre. *The Life of a South African Tribe*. London: MacMillan, 2nd rev. ed., 1927, pp. 449-83.

1326 LEE, Richard B. "Kung Bushman Violence." Paper presented at the 68th annual meeting of the American Anthropological Association, New Orleans, 1969.

1327 LINTON, Ralph. *The Tanala: A Hill Tribe of Madagascar.* Chicago: Field Museum of Natural History, 1933, pp. 248-51.

1328 MORRIS, Donald P. *The Washing of the Spears: A History of the Rise of the Zulu Nation under Shaka and Its Fall in the Zulu War of 1879.* New York: Simon and Schuster, 1965.

1329 OMER-COOPER, J.D. *The Zulu Aftermath: A 19th Century Revolution in Bantu Africa.* Evanston, Illinois: Northwestern University Press, 1966.

1330 ——. "Shaka and the Rise of the Zulu." *Tarikh,* (1965), 1:30-41.

1331 OTTERBEIN, Keith. "The Evolution of Zulu Warfare." *Kansas Journal of Sociology,* (1964), 1:27-35.

1332 READ, Margaret. "The Moral Code of the Ngoni and Their Former Military State." *Africa,* (1938), 11:1.

1333 ——. *The Ngoni of Nyasaland.* London: Oxford University Press, 1960, pp. 31-43.

1334 RITTER, E.A. *Shaka Zulu: The Rise of the Zulu Empire.* New York: G.P. Putnam's Sons, 1957.

1335 SMITH, Edwin W. *The Ila-speaking Peoples of Northern Rhodesia.* v. 1. London: Macmillan, 1920, pp. 175-78, 278-79.

T / MIDDLE EAST (including North Africa)

1336 BARTH, Frederick. *Principles of Social Organization in Southern Kurdistan.* Oslo: Brodrene Jorgensen Boktr., 1953, pp. 73-77, 117, 138-43.

1337 BAUER, I. "Technica y tactica de la pirateria Berberisca." *Africa: revista de accion española,* (1950), 7:442-44.

1338 BLACKMAN, Winifred S. *The Fellahin of Upper Egypt.* London: Harrap, 1927, chapt. 8, "Inter-village Fights: The Law of Revenge."

1339 BLACKWELL, H.F., ed. *The Occupation of Hausaland 1900-1904.* New York: Barnes and Noble, 1927.

1340 COON, Carleton S. *Caravan: The Story of the Middle East.* New York: Holt, 1952, pp. 205, 208-9, 218-19, 315-17, 332-33.

1341 DICKSON, Harold R.P. *The Arab of the Desert.* London: Allen and Unwin, 1951, chapt. 26, "Badawin Warfare."

1342 DRAKE-BROCKMAN, Ralph. *British Somaliland.* London: Hurst and Blackett, 1912, pp. 145, 149-55.

1343 DUMEZIL, Georges. "De Quelques faux massacres." *Turk Antropologii Mecmuasi,* (1927), 4:39-46.

1344 DUNDES, Alan, and Bora Ozok. "Penis and Anus in the Strategy of Turkish Male Verbal Dueling Rhymes." Paper presented at the 67th annual meeting of the American Anthropological Association, Seattle, 1968.

1345 EDMONDS, C.J. "Luristan: Pish-i-Kuh and Bala Gariveh." *Geographical Journal,* (1922), 59:335-56, 437-53.

1346 FAKHOURI, Hani. "Tribal Law and the Settlement of Disputes in Jordan." Paper presented at the 69th annual meeting of the American Anthropological Association, San Diego, 1970.

1347 GLUBB, John Bagot. "Arab Chivalry." *Royal Central Asian Society, Journal,* (1951), 5:284-302.

1348 HART, David Montgomery. "An Ethnographic Survey of the Riffian Tribe of Aith Wuryaghil." *Tamuda,* (1954), 2:55-86.

1349 HASSAN, Alhaji. *A Chronicle of Abuja.* (trans. by Frank Heath.) Ibadan: Ibada University Press, 1952, pp. 17-28.

1350 ———, and H.J. Liebesny, eds. *Law in the Middle East: v. 1. Origin and Development of Islamic Law.* Washington: Middle East Institute, 1955, pp. 355, 358, 367-70.

1351 KHADDURI, Majid. *War and Peace in the Law of Islam.* Baltimore: Johns Hopkins University Press, 1955.

1352 LEACH, Edmund Ronald. *Social and Economic Organization of the Rowanduz Kurds.* London: London School of Economics and Political Science, 1940, chapt. 5, "Warfare."

1353 LEWIS, Joan M. "Clanship and Contract in Northern Somaliland." *Africa,* (1959), 29:274-93.

1354 ———. *Peoples of the Horn of Africa.* London: International African Institute, 1955, pp. 107-10.

1355 LEWIS, William H. "Feuding and Social Change in Morocco." *JCR,* (1961), 5:43-54.

1356 MASTERS, William M. *Rowanduz: A Kurdish Administrative and Mercantile Center.* Ann Arbor: University Microfilms, no. 7689,

University of Michigan,   1953, pp. 4, 28, 36, 103, 180-87, 242, 328, 336-39.

1357 MUFFETT, David J. *Concerning Brave Captains: Being a History of the British Occupation of Kano and Soloto and the Last Stand of the Fulani Forces.* Ontario: Don Mills, 1964.

1358 MUSIL, Alois. *The Manners and Customs of the Rwala Bedouins.* New York: American Geographical Society, 1928.

1359 OMAN, G. "Notizie sulla terminologia militare in alcuni paesi arabi." Rome, *Oriente Moderno,* (1962), 43:278-84.

1360 ORAL, Z. "Karamanogullari Zamaninda Bir Sinir Ihtil afi (Folklore)." *Turk Etnografya dergisi,* (Ankara, 1957), 2:67-68.

1361 PETRASCH, E. "So zog der Turk ins Feld . . ." *Atlantis,* (1957), 29:555-66.

1362 RASWAN, Carl R. *Black Tents of Arabia.* New York: Creative Age Press, 1947, pp. 3, 14-15, 38-46, 86, 97-100.

1363 ROTHENBERGER, John E. "Conflict and Conflict Resolution in a Sunni Muslim Village in Lebanon." Paper presented at the 67th annual meeting of the American Anthropological Association, Seattle, 1968.

1364 SLATIN, Rudolf C. *Fire and Sword in the Sudan: A Personal Narrative of the Fighting and Serving the Dervishes, 1879-1895.* London, 1896.

1365 SWEET, L.E. "Camel Raiding of North Arabian Bedouin: A Mechanism of Ecological Adaptation." AA, (1965), 67:1132-50.

1366 WALDMAN, M.R. "A Note on the Ethnic Interpretation of the Fulani Jihad." *Africa,* (London, 1966), 36:286-91.

T.1 / Egypt & Mesopotamia

1367 BENEDICT, Marian J. *The God of the Old Testament in Relation to War.* New York, 1927.

1368 BREASTED, J.H. *The Battle of Kadesh.* Chicago: University of Chicago Press, 1903.

1369 BRUNET, A.M. "La Guerre dans la Bible." Lyon, *Lumière et vie,* (1958), 38:31-47.

1370 BURNE, Alfred H. "The Battle of Kadesh, 1288 B.C." In *The Art of War on Land.* Harrisburg, Pennsylvania: Military Service Publishing Co., 1947, pp. 37-46.

1371 DIVALE, W.T. "The Evolution of the Chiefdom in the Ancient Near East." Paper presented at the annual meeting of the American Folklore Society, Los Angeles, 1970.

1372 FAULKNER, R.O. "The Euphrates Campaigns of Thutmosis III." JEA, (1946), 32:39-42.

1373 ———. "The Battle of Megiddo." JEA, (1942), 28:2-15.

1374 KANTOR, H.J. "The Early Relations of Egypt with Asia." JNES, (1942), 1:174-213.

1375 KUENTZ, C. "La Bataille de Qadech." *Memoires de l'Institut Français d'Archéologie Arientale du Caire,* (1928-34), v. 55.

1376 MALAMAT, A. "Military Rationing in Papyrus Anastasi I and the Bible." *Mélanges bibliques vediges en l'honneur d'André Robert,* (Paris), 1957.

1377 MODI, J.J. "The Bombardment of Paris from a Distance of 60 Miles, Supposed to be a Mystery. An Instance of a Somewhat Similar Mystery in the Ancient History of Persia." JASB, (1918), 11:389-96.

1378 NELSON, Harold H. "The Naval Battle Pictured at Medinet Habu." JNES, (1943), 2:40-55.

1379 ———. *The Battle of Megiddo.* Chicago: Ph.D. thesis, University of Chicago, 1913. Private ed. distributed by the University of Chicago library.

1380 PRITCHARD, J.B. "Civil Defense at Gibeon." *Expedition,* (1962), 5:10-17.

1381 SCHULMAN, A.R. "Siege Warfare in Pharaonic Egypt." NA, (1964), 73:12-21.

1382 TOLKOWSKY, S. "Gideon's Three Hundred." *Journal of the Palestine Oriental Society,* (1925), 5:69-74.

1383 VAUX, R. "Les Combats singuliers dans l'ancien Testament." *Biblica,* (1959), 40:495-508.

1384 WARDMAN, A.E. "Tactics and the Tradition of the Persian Wars." Wiesbaden, *Historia,* (1959), 8:49-60.

1385 WINLOCK, H.E. *The Slain Soldiers of Neb-hepet-Re Mentuhotep.* New York, 1945.

1386 WOOLLEY, C.L. *The Town Defenses (of Carchemish).* v. 3. London, 1921.

1387 YADIN, Yigael. *The Art of Warfare in Biblical Lands.* London: Weienfeld and Nicolson, 1963.

1388 ——. *The Scroll of the Way of the Sons of Light against the Sons of Darkness.* Oxford: Oxford University Press, 1962.

1389 ——. "Solomon's City Wall and Gate at Gezer." IEJ, (1958), 8:80-86.

1390 ——. "Some Aspects of the Strategy of Ahab and David." *Biblica,* (1955), 36:332-51.

1391 ——. "The Earliest Record of Egypt's Military Penetration into Asia?" IEJ, (1955), 5:1-16.

1392 ——. "The Blind and the Lame and David's Conquest of Jerusalem." *Proceedings of the First World Congress of Jewish Studies, Summer 1947.* Jerusalem, 1952.

1393 YEIVIN, S. "Canaanite and Hittite Strategy in the Second Half of the Second Millennium B.C." JNES, (1950), 9:101-7.

U / EUROPE & THE SOVIET UNION

1394 ADCOCK, Frank E. *Roman Art of War under the Republic.* Scranton: Barnes and Noble, 1971.

1395 ALLEN, W.E.D. "The Caucasus." In John Buchan, ed., *The Baltic and Caucasian States.* London: Hodder and Stoughton, 1923, pp. 167-269.

1396 ALLEN, William Edward D. *A History of the Georgian People from the Beginning down to the Russian Conquest in the Nineteenth Century.* London: K. Paul, Trench, Trubner, 1932, pp. 275-81.

1397 BALIKCI, A. "Quarrels in a Balkan Village." AA, (1965), 67:1456-69.

1398 BLOK, A. "Mafia and Peasant Rebellion as Contrasting Factors in Sicilian Latifandism." *Archives Européennes de Sociologie,* (1969), 10:95-116.

1399 BULLOUGH, Vern L. "The Roman Empire vs. Persia, 363-502: A Study of Successful Deterrence." JCR, (1963), 7:55-68.

1400 CAESAR, Caius Julius. *Alexandrian, African and Spanish Wars.* Trans. by A.G. Way. Cambridge: Harvard University Press, 1955.

1401 ——. *The Gallic War.* Trans. by H.J. Edwards. London: W. Heinemann; New York: Putnam's Sons, 1930.

1402 ——. *The Civil Wars.* Trans. by A.G. Peskett. London: W. Heinemann; New York: Putnam's Sons, 1921.

1403 COCKLE, Maurice J.D. *A Bibliography of English Military Books up to 1642 and of Contemporary Foreign Works.* London: Simpkin, Marshall, Hamilton, Kent and Co., 1900.

1404 FOURDRIGNIER, E. "Les Chars de guerre au second age du fer." *La Revue Prehistorique,* (1906), 1:52-65, 73-82.

1405 GARDNER, E.A. "War, War-Gods (Greek and Roman)." In *Hasting's Encyclopedia of Religion and Ethics,* v. 12. Edinburgh, 1921.

1406 GARNETT, Lucy Mary J. "Albanian Women." *The Women of Turkey and Their Folk-lore,* v. 2. London: David Nutt, 1891, pp. 221-35, 270, 273, 280.

1407 GEIGER, Bernard, et al. *The Caucasus,* 2 vols. New York: Columbia University, 1956, pp. 249, 544-48.

1408 GJESSING, Gutorm. *Changing Lapps.* London: Dept. of Anthropology, London School of Economics and Political Science, 1954, pp. 47-50, 59-62.

1409 GRIGOLIA, Alexander. *Custom and Justice in the Caucasus: The Georgian Highlanders.* Philadelphia, 1939, pp. 67-68, 86, 143-45.

1410 HASLUCK, Margaret. *The Unwritten Law in Albania.* Cambridge: University Press, 1954, pp. xii-xv, 43-44, 74-81, 85, 93-94, 99-103, 108-9, 125-27, 134, 144, 152, 158-59, 162, 212-18, 252, 263-65, 268, 272.

1411 HERRMANN, J. "Die vor und frühgeschtlichen Wehranlogen Gross-Berlins und des Bezirkes Potsdam." Berlin: *Ethnographisch-archaologische zeitschrift,* (1960), 1:38-41.

1412 HOMER. *The Iliad.* Trans. by Richard Lattimore. Chicago: University of Chicago Press, 1961.

1413 LANE, Rose (Wilder). *Peaks of Shala.* New York: Harper, 1923, pp. 11-12, 28-31, 41-42, 50-51, 149-50, 271-79.

1414 LUZBETAK, Louis J. *Marriage and the Family in Caucasia.* Vienna-Modling: St. Gabriel's Mission Press, 1951, pp. 56, 58, 70, 101, 150-51, 160-61, 168-69, 186, 194.

1415 MEIGS, Reveril. "Some Geographical Factors in the Peloponnesian War." *Geographical Review.* (1961), 51:370-80.

1416 OMAN, Charles W.C. *The Art of War in the Middle Ages, A.D. 378-1515.* 2 vols. Ithaca: Great Seal Books, 1960. (original edition, 1898.)

1417 ——. *A History of the Art of War in the Sixteenth Century.* New York: Dutton, 1937.

1418 PAINE, Robert. *Coast Lapp Society, I: A Study of Neighbourhood in Revsbotn Fjord.* Tromso, Norway, 1957, pp. 229, 241-43, 286-87.

1419 PLUTARCH. *The Lives of the Noble Grecians and Romans.* (Dryden translation) rev. ed. by Arthur H. Clough. New York: Modern Library, 1936.

1420 REDLICH, Marcellus D. *Albania Yesterday and Today.* Worcester: Albanian Messenger, 1936, pp. 15, 23, 31, 68-70, 155-56, 202, 282.

1421 SCHRENCK, Leopold von. *Die Völker des Amur-Landes.* St. Petersburg: Kaiserliche Akademie der Wissenschaften, 1881-1895, pp. 427, 554-57, 574, 634-35, 771-72.

1422 SHTERNBERG, Ler Iakovlevich. *Galiaki, orochi, gol'dy, negidal'tsy ainy; stat'i i materialy.* Khabarorsk: Dal'giz, 1933, pp. 59, 94-105, 117, 127, 275, 286, 338, 355-58, 383.

1423 THOMPSON, E.A. "Early Germanic Warfare." *Past and Present,* (1958), 14:2-29.

1424 THUCYDIDES. *The Peloponnesian War.* New York: Modern Library, 1934.

1425 VERNANT, J.P., ed. *Problèmes de la guerre en Grèce ancienne.* La Haye: Mouton, 1968.

1426 VOGT, J. "Struktur der antiken Sklavenkriege." Mainz, *Akademie der Wissenschaften und der Literatur. Abhandlungen der geistesund sozialwissenschaftlichen Klasse,* (1957), 1:7-57.

V / ASIA & INDIA

1427 AYYANGAR, Sri M.A. "Conflicts and Tensions in the National Perspective." CTCT, (1969), 15-24.

1428 BASCOM, W.R. "Ponape: The Tradition of Retaliation." *Far East Quarterly,* (1950), 10:56-62.

1429 BATCHELOR, John. *Ainu Life and Lore; Echoes of a Departing Race.* Tokyo: Kyobunkwan, 1927, pp. 287-89, 415-21.

1430 BHASKARAN, K. "Social Tension and Conflicts in Contemporary Indian Society from a Psychiatric Point of View." CTCT, (1969), 45-51.

1431 BURMAN, B.K. Roy. "Conflict and Tension in Rural India." CTCT, (1969), 201-6.

1432 CHAKRAVARTI, P.C. *The Art of War in Ancient India.* Dacca, (n.d.)

1433 DARLING, Malcolm. *Wisdom and Waste in the Punjab Village.* London, New York: Oxford University Press, 1934, pp. 45-48.

1434 DATE, G.T. *The Art of Warfare in Ancient India.* London, 1929.

1435 de BEAUCLAII, Inez. "Fighting and Weapons of the Yami of Botel Tobago." Taiwan, *Bulletin of the Institute of Ethnology, Academia Sinica,* (Spring 1958), 5:87-111.

1436 DIKSHITAR, V.R. Ramachandra. *War in Ancient India.* Madras: Macmillan & Co., Ltd., 1944.

1437 DUBOIS, J.A. *Hindu Manners, Customs and Ceremonies.* Oxford: Clarendon Press, 1906.

1438 EBERHARD, Wolfram. *Conquerors and Rulers: Social Forces in Medieval China.* Leiden: Brill, 1952.

1439 EKVALL, R.B. "Peace and War among the Tibetan Nomads." AA, (1964), 66:1119-48.

1440 ———. "Nomadic Pattern of Living among the Tibetans as Preparation for War." AA, (1961), 63:1250-63.

1441 ELWIN, Verrier, ed. *India's North-east Frontier in the Nineteenth Century.* London: Oxford University Press, 1959.

1442 ———. *Maria Murder and Suicide.* Oxford: Oxford University Press, 1943.

1443 FEINGOLD, David. "Consensual Settlement in Akha Law." (Northern Thailand). Paper presented at the 69th annual meeting of the American Anthropological Association, San Diego, 1970.

1444 FISHER, Margaret; L. Rose; and R. Huttenback. *Himalayan Battleground: Sino-Indian Rivalry in Ladakh.* New York: Praeger, 1963.

1445 FURER-HAIMENDORF, Christoph von. *Himalayan Barbary.* London, 1955.

1446 GARDNER, G.B. *Keris and Other Malay Weapons.* Singapore: Progressive Publishing Co., 1936, pp. 118-32.

1447 GOMPERTZ, M. "The Human Side of the Indian Army." *Asia,* (1931), 33:186-90.

1448 GULLICK, J.M. "Malayan Warfare." *Malaya in History,* (1957), 3:116-19.

1449 HITCHCOCK, J.T. "The Idea of the Martial Rajput." JAF, (1958), 71:216-23.

1450 HODSON, T.C. "Head-hunting among the Hull Tribes of Assam." *Folklore,* (1909), 20:132-143.

1451 KENN, Charles W. "The Army and Navy of Kamehameha I." *U.S. Naval Institute Proceedings,* (1943), 71:(11) 1335-1339.

1452 LATTIMORE, Owen. "China and Barbarians." In *Empire in the East,* J. Barnes, ed. Garden City, 1934, pp. 3-36.

1453 ———. *Inner Asian Frontiers of China.* New York: American Geographical Society, 1940.

1454 ———. "Origins of the Great Wall of China." *Geographical Review,* (1937), 27:529-49.

1455 LEACH, Edmund. *Political Systems of Highland Burma.* Cambridge: Harvard University Press, 1954, pp. 89-98.

1456 LIN, Yueh-hwa. *Liang-shan I-chia.* (The Lolo of Liang-shan; HRAF translation) Shanghai: Commercial Press, 1947, pp. 81-109.

1457 LITVINSKIJ, D.A., and I.V. P'Jankov. "Voennoe delo u narodor srednej Azii v VI-IV vv. do u. e." Moscow, *Vestnik drevnej Istorii,* (1966), 3:36-52.

1458 MARSHALL, Harry Ignatius. *The Karen People of Burma.* Columbus: University of Ohio Press, 1922, pp. 152-60, "Warfare and Weapons."

1459 MAYBON, Charles B. *Histoire moderne du pays d'Annam 1592-1820.* (HRAF translation) Paris: Typographie Plon-Nourrit, 1919, pp. 383-429.

1460 McGOVERN, Mrs. J.B.M. *Among the Head-hunters of Formosa.* London, 1922.

1461 MEHTA, B.H. "Forestry and Tensions in Tribal Areas." CTCT, (1969), 154-70.

1462 MILLS, J.P. "The Effect on the Tribes at the Naga Hills District of Contacts with Civilization." Calcutta, *Census of India 1931,* 3: App. II-IV.

1463 ———. "Head-hunting in Assam." *Asia,* (1926), 26:876-883, 904-907.

1464 MODI, J.J. "The Aghukhoh of Sema Negas of the Assam Hills and the Chah of Kabulia." JASB, (1923), 12:609-16.

1465 MUKHERJEE, P.N. "The Importance of Peace in Ancient Indian Culture." Agra, *Agra University Journal of Research (letters),* (1957), 5:199-203.

1466 NANDI, Santibhushan. "Resolution of Conflicts in a Bengal Village." CTCT, (1969), 264-73.

1467 ORENSTEIN, Henry. "Types of Conflict: A Case Study of a Region in India." Paper presented at the 62nd annual meeting of the American Anthropological Association, San Francisco, 1963.

1468 ORUI, S. "Ancient Battle in Japan." (text in Japanese) *Kokogaku zasshi*, (1916), 6(5).

1469 PEARN, Bertie R. *Burma Background*. London: Longmans; Calcutta: Green, 1943, pp. 8-28.

1470 POLLARD, Samuel. *In Unknown China*. Philadelphia: Lippincott, 1921, pp. 39-47, 196-203, 239-44, 307-11.

1471 RADCLIFFE-BROWN, A.R. "Andaman Peacemaking." PH, (1953), 532-33.

1472 RAO, G.R.S. "Social Change and Conflicts in Tribal Communities." CTCT, (1969), 181-91.

1473 RASTOGI, P.M. "Factions in Kurmipur." *Man in India*, (1965), 45:289-95.

1474 RAY, H.C. "Notes on War in Ancient India." *Calcutta University, Journal of the Department of Letters*, (1927), 7:1-80.

1475 RAY, P.C. "Stereotypes and Tensions among the Muslims and Hindus in a Village in Bengal." CTCT, (1969), 228-53.

1476 ROBINSON, Marguerite S. "Sorcery and Social Change in a Ceylon Colonization Scheme." Paper presented at the 68th annual meeting of the American Anthropological Association, New Orleans, 1969.

1477 SACHACHERMEYR, F. "Streitwagen und Streitwagenbild im Alten Orient und bei den mykenischen Griechan." *Anthropos*, (1951), 46:705-53.

1478 SACHCHIDANANDA. "Caste and Conflict in a Bihar Village." CTCT, (1969), 254-63.

1479 SAHAY, K.N. "Christianity as a Factor of Tension and Conflict among the Tribals of Chotanagpur." CTCT, (1969), 274-98.

1480 SARAN, A.B. "Social Conflict and Tribal Tension." CTCT, (1969), 134-45.

1481 SEN, Sachin. "Aspects of Conflicts and Tensions in Contemporary Indian Society." CTCT, (1969), 25-33.

1482 SINGH, S.D. *Ancient Indian Warfare with Special Reference to the Vedic Period*. Leiden: B.J. Brill, 1965.

1483 SINHA, B.P. "Art of War in Ancient India, 600 B.C.-A.D. 300." Paris, *Cahiers d'histoire mondiale*, (1957), 4:123-60.

1484 SRIVASTAVA, A.L. "Review of Bajwa, 1964, *Military . . ." Journal of Indian History*, (1964), 43:578-80.

1485 SRIVASTAVA, L.R.N. "Inter-village and Intra-village Conflict in a Tribal Society." CTCT, (1969), 171-80.

1486 VIAL, L.G. "Knights of the Stone Age." *Asia*, (1939), 39:408-12.

1487 VIDYARTHI, L.P., ed. *Conflict Tension and Cultural Trends in India.* Calcutta: Punthi Pustuk, 1969.

1488 ———. "Inter-group Conflicts in India: Tribal, Rural and Industrial." CTCT, (1969), 34-44.

1489 VILINBAHOV, V.B. "Kistorii vlijanija kabardincer na voennyi byt kazacestra." Nalcik, *Ucenye Zapiski (Kabardinskij-Balkarskij Naucny j Institut). Serija Ystorii i Filologii,* (1965), 23:125-32.

1490 WALES, Horace G.Q. *Ancient South-East Asian Warfare.* London: B. Quaritch, 1952.

1491 WANDREKAR, D.N. "Tension and Conflicts in the Tribal Regions of Contemporary India." CTCT, (1969), 124-33.

1492 WIENS, Harold J. *China's March Toward the Tropics.* Hamden, Connecticut: Shoe String Press, 1954, pp. 288-93, 102-7.

1493 WIGRAM, Sir Kenneth. "Defense in the N.W. Frontier Province." *Journal of the Royal Central Asian Society,* London, (Jan. 1937), 24.

1494 WILDEN, G.A. *Handleiding voor de Vergelijkende Volkenkunde van Nederlandsch-Indie.* (HRAF translation) Leiden: Brill, 1893, Chapt. 16, "Warfare."

1495 WOODTHROPE, R.G. *The Lushai Expedition, 1871-72.* London, 1875.

1496 YADARA, J.S. "Factionalism in a Haryana Village." AA, (1968), 70:898-910.

W / OCEANIA

W.1 / Philippines & Indonesia

1497 BARTON, Roy Franklin. *The Religion of the Ifugaos.* Menasha: American Anthropological Association, 1946, pp. 48, 140-44, 156-57.

1498 ———. *Philippine Pagans: The Autobiographies of Three Ifugaos.* London: Routledge, 1938, pp. 6-7, 52-53, 242, 261.

1499 ——. *The Half-way Sun: Life among the Headhunters of the Philippines.* New York: Brewer and Warren, 1930, pp. 156-64, 302-3.

1500 ——. *Ifugao Law.* Berkeley: University of California Press, 1919, pp. 36, 75-78, 99-100, 106-9, 152.

1501 CHABOT, Hendrik T. *Verwantschap, stand en sexe in Zuild-Celebes.* Groningen: J.B. Wolters, 1950, pp. 68-69, 91-92.

1502 DOWNS, R.E. "Criticism? A Rejoinder to Professor Fischer." BTLV, (1955), 3:280-85.

1503 DUBOIS, Cora. *The People of Alor: A Social-Psychological Study of an East Indian Island.* Minneapolis: University of Minnesota Press, 1944, pp. 127-29, 198-99.

1504 ——. "Why People Quarrel in Alor." *Asia,* (1941), XLI:91-94.

1505 GOMES, Edwin H. *Seventeen Years among the Sea Dyaks of Borneo.* London: Seeley, 1911, pp. 23-24, 81-83, 136, 142.

1506 HART, Donn Vorhis. *Barrio Caticugan: A Visayan Filipino Community.* Syracuse University, Ph.D. thesis, 1954, pp. 162, 165-66, 234-39, 254-57.

1507 HOSE, Charles, and W. McDougall. *The Pagan Tribes of Borneo.* 2 vols. New York: Barnes and Noble, 1966.

1508 JENKS, Albert E. *The Bantoc Igorot.* Manila: Bureau of Public Printing, 1905, pp. 170, 177-80, 192.

1509 KENNEDY, Raymond. *Field Notes on Indonesia: Ambon and Ceram, 1949-50.* New Haven: Human Relations Area Files, unpublished manuscript, pp. 292-97, 306-7.

1510 ——. *The Ageless Indies.* New York: John Day, 1942, pp. 100-5, 166-67.

1511 KIEFER, Thomas M. *Tausug Armed Conflict: The Social Organization of Military Activity in a Philippine Moslem Society.* Chicago: University of Chicago, Philippine Studies Program, Research Series No. 7, 1969.

1512 ——. "Modes of Social Action in Armed Combat: Affect, Tradition and Reason in Tausug Private Warfare." Paper presented at the 68th annual meeting of the American Anthropological Association, New Orleans, 1969.

1513 ——. "Institutionalized Friendship and Warfare among the Tausug of Jolo." *Ethnology,* (1968), 7:225-44.

1514 KROEBER, Alfred L. *Peoples of the Philippines.* New York: American Museum of Natural History, 2nd rev. ed., 1928, pp. 167, 174.

1515 LAMBRECHT, Francis. *The Mayawyaw Ritual.* Washington, D.C.: Catholic Anthropological Conference, 1932, pp. 445-60.

1516 LOW, Hugh. *Sarawak, Its Inhabitants and Productions.* London: Richard Bentley, 1848, pp. 212, 218-25.

1517 ROTH, H. Ling, ed. *The Natives of Borneo.* JAI, (1893), 22:22-64.

1518 STOCKDALE, J.J. *Sketches, Civil and Military, of the Island of Java.* London, 1812.

1519 SWELLENGREBEL, J.L. *Bali: Studies in Life, Thought, and Ritual.* The Hague: W. van Hoeve, 1960, pp. 194-97, 362-68.

1520 VANOVERBERGH, Morice. *The Isneg Life Cycle.* Washington, D.C.: Catholic University Press, 1936-1938, pp. 144-47, 156-57, 175-76, 223-24.

1521 VAYDA, A.P. "The Study of the Causes of War, with Special Reference to Head-hunting Raids in Borneo." *Ethnohistory,* (1969), 16:211-24.

1522 WILKEN, G.A. *Handleiding voor de Volkenkunde van Nederlandsche-Indie.* Leiden: Brill, 1893, pp. 389-92, 467.

1523 WILSON, Laurence L. *Apayao Life and Legends.* Baguio, P. I, 1947, pp. 3, 14-15, 48-49, 54-60, 158-60.

W.2 / Melanesia, Polynesia, & Micronesia

1524 ALLARDYCE, W.L. "The Fijians in Peace and War." *Man,* (1904), 4:69-73.

1525 BARKER, George T. "A War on the Nakauvadra." *Fijian Society Transactions for 1924,* (1925), 37-40.

1526 BEAUCLAIR, I. "Fightings and Weapons of the Yami of Botel Tobago." (Melanesia) *Bulletin of the Institute of Ethnology, Academia Sinica,* 5:87-111.

1527 BELL, F.L. "Warfare Among the Tanga." *Oceania,* (1935), 5:253-79.

1528 BLACKWOOD, Beatrice. *Both Sides of Buka Passage.* Oxford: Clarendon Press, 1935, pp. 192, 469-70, 503.

1529 DERRICK, R.A. "Fijian Warfare." *Fiji Society of Science and Industry Transactions and Proceedings*, (1953), 2:137-46.

1530 ECKERT, G. "Die Kopfjagd im Caucatal." ZFE, (1939), 71:305-18.

1531 EMORY, K.P. "Warfare." *Ancient Hawaiian Civilization.* Honolulu: Kamehameha Schools, 1939, pp. 229-36.

1532 FINSCH, Otto. "Kriegsfuhrung auf den Marshall-Inseln." *Die Neue Gartenlaube,* (1882), 29:700-3.

1533 ———. "Bilder aus dem stillen Ocean: 1. Kriegsfuhrung auf den Marshall-Inseln" *Gartenlaube,* (1881), 29:700-3.

1534 FIRTH, Raymond W. *Social Change in Tikopia.* London: Allen and Unwin, 1959, pp. 210, 305.

1535 FORTUNE, R.F. *Manus Religion.* Philadelphia: American Philosophical Society, 1935, pp. 40, 124, 213, 235-36, 252-53, 282, 345, 365.

1536 GLADWIN, Thomas, and S.B. Sarason. *Truk: Man in Paradise.* New York: Wenner-Gren Foundation for Anthropological Research, 1953, pp. 31-32, 36, 40-42, 47, 50, 64, 67-70, 111-12, 139-41, 183, 278.

1537 GOLDMAN, Irving. "Status Rivalry and Cultural Evolution in Polynesia." AA, (1955), 57:680.

1538 GOODENOUGH, Ward H. *Kin and Community on Truk.* New Haven: Yale University Press, 1951, pp. 53-54, 84-86, 141, 144, 148-60, 173-83.

1539 GRAEBNER, Fritz. "Völkerkunde der Santa-Cruz-Inseln." *Ethnologica,* (1909), 1:71-184.

1540 HAAS, Johann Gustav (Salesius, Van P.). *Die Karolinen-Insel Yap.* Berlin: W. Susserott, 1907, pp. 57, 64, 66, 69, 79, 110-11, 117-18, 120, 126-27, 130.

1541 HADDON, A.C., and A. Wilkin. "Warfare." RCAETS, (1904), 5:298-307.

1542 HANDY, Edward Smith C. *The Native Culture of the Marquesas.* Honolulu: Bernice P. Bishop Museum, 1923, pp. 56, 123-24, 131-37.

1543 HEINE-GELDERN, R. "Politische Zweiteilung, Exogamie und Kriegsursachen auf der Osterinsel." *Ethnologica,* (1960), 2:241-73.

1544 HOCART, A.M. "Warfare in Eddystone of the Solomon Islands." JAI, (1931), 61:301-24.

1545 ———. *Lau Islands, Fiji.* Honolulu: Bernice P. Bishop Museum, 1929, pp. 141-44.

1546 KNIBBS, Stanley G.C. *The Savage Solomons as They Were and Are, a Record of a Head-hunting People. . . .* Philadelphia: Lippincott, 1929.

1547 KRAMER, Augustin. *Ergebnisse der Sudsee-Expedition 1908-1910, II. Ethnographie: B. Mikronesien, v. 5.* Hamburg: Friederichsen, de Gruyter, 1932, pp. 12-16, 69-73, 79, 87, 268-70.

1548 ——, and H. Nevermann. *Ralik-Ratak (Marshall-Inseln).* Hamburg: Friederichsen, de Gruyter, 1938, pp. 6, 201-4.

1549 LAYARD, John W. *Stone Men of Malekula.* London: Chatto and Windus, 1942, pp. 595-603.

1550 LINTON, Ralph. "Marquesan Culture." In Abram Kardiner, *The Individual and His Society.* New York: Columbia University Press, 1939, pp. 138-96.

1551 LORIA, Lamberto. "Notes on the Ancient War Customs of the Natives of Logea and Neighborhood." *British New Guinea Annual Report, 1894-5,* (1896), pp. 44-54.

1552 MALINOWSKI, Bronislaw. *Crime and Custom in Savage Society.* New York: Harcourt, Brace, 1926.

1553 ——. *Argonauts of the Western Pacific.* London: George Routledge, 1922, pp. 291-96, 406.

1554 ——. "War and Weapons among the Natives of the Trobriand Islands." *Man,* (1920), 20:10-12.

1555 MEAD, Margaret. *New Lives for Old: Cultural Transformation—Manus, 1928-1953.* New York: Morrow, 1956, pp. 26, 33, 52, 71-73, 89, 131, 143, 263, 318, 401.

1556 ——. *Growing Up in New Guinea.* New York: Morrow, 1930, pp. 58, 114, 194-95, 302.

1557 ——. "Melanesian Middlemen." *Natural History,* (1930), 30:115-30.

1558 METRAUX, Alfred. *Ethnology of Easter Island.* Honolulu: Bernice P. Bishop Museum, 1940, pp. 147-50.

1559 MULLER, Wilheim. *Yap. Band 2, Halbband 1.* Hamburg: Friederichsen, 1917, pp. 9, 28, 188-93, 258-66, 277.

1560 PARKINSON, R. *Zur Ethnographie der Nordwest lichen Salomo-Inseln.* Dresden: Konigliche Zoologisch und Anthropologisch-Ethnographisch Museum, 1899, pp. 7, 25.

1561 PORTER, David. *A Voyage in the South Seas in the Years 1812, 1813, and 1814.* London: Sir Richard Phillips, 1823, pp. 86-88.

1562 POWDERMAKER, Hortense. *Life in Lesu: The Study of a Melanesian Society in New Ireland.* New York: Norton, 1933, pp. 298, 304, 341-45.

1563 POWELL, H.A. "Competitive Leadership in Trobriand Political Organization." JAI, (1960), 90:118-45.

1564 ROHEIM, Geza. "Yaboaine, a War God of Normansby Island." *Oceania*, (1946), 4:319-36.

1565 ROMANUCCI SCHWARTZ, L. "Conflits fonciers à Mokerang, village Matankov des Îles de l'Amirauté." *Homme*, (1966), 6:32-52.

1566 SCHEFFLER, H.W. "The Genesis and Repression of Conflict: Choiseul Island." AA, (1964), 66:789-804.

1567 SELIGMANN, Charles Gabriel. "The Northern Massim." *The Melanesians of British New Guinea.* Cambridge: University Press, 1910, pp. 663-69, 698.

1568 SENFFT, Arno. "Ethnographische Beiträge über die Karolineninsel Yap." *Petermanns Mitteilungen*, (1903), 49:49-60, 83-87.

1569 ———. "Die Marshall-Insulaner." In R.S. Steinmetz, ed., *Rechtsverhaltnisse von eingeborenen Völkern in Afrika und Ozeanien.* Berlin: Julius Springer, 1903, pp. 425-55.

1570 SPEISER, Felix. "Völkerkundliches von den Santa-Cruz-Inseln." *Ethnologica*, (1916), 2:153-214.

1571 STAIR, John Bettridge. *Old Samoa; Or, Flotsam and Jetsam from the Pacific Ocean.* London: The Religious Tract Society, 1897, pp. 98-101, 126, 245-47, 250-58.

1572 THOMPSON, Laura. *Southern Lau, Fiji: An Ethnography.* Honolulu: Bernice P. Bishop Museum, 1940, pp. 103-5, 110.

1573 TIPPETT, Rev. A.P. "The Nature and Social Functions of Fijian War (focus . . . 1839-1846)." *Transactions and Proceedings of the Fiji Society*, (1958), 5:137-155.

1574 TODD, J.A. "Redress of Wrongs in South-West New Britain." *Oceania*, (1936), 6:401-40.

1575 TURNER, George. *Samoa, A Hundred Years Ago and Long Before.* London: Macmillan, 1884, pp. 23, 57-58, 65-66, 189-96.

1576 WEDGWOOD, C.H. "Some Aspects of Warfare in Melanesia." *Oceania*, (1930), 1:5-33.

1577 WILKIN, A. "Tales of the Warpath." RCAETS, (1904), 5:308-19.

W.3 / New Guinea

1578 AUSTEN, Leo. "Karigara Customs." (War) *Man*, (1923), 23:35-36.

1579 BAREREBA, S. "How My Grandfather Killed Mr. I. Green." *South Pacific*, (1959), 10:129-32.

1580 BERNDT, R.M. "Warfare in the New Guinea Highlands." AA, (1964), 66:183-203.

1581 ———. "Interdependence and Conflict in the East Central Highlands of New Guinea." *Man*, (1955), 55:105-7.

1582 BROWN, Paula. "From Anarchy to Satrapy." AA, (1963), 65:1-15.

1583 ———. "Chimbu Tribes: Political Organization in the E. Highlands of New Guinea." SWJA, (1960), 16:22-35.

1584 DIVALE, W.T. "Kapauku Warfare, Calories, and Population Control." Paper presented at the annual meeting of the American Anthropological Association, New York, 1971.

1585 FORTUNE, R.F. "New Guinea Warfare: Correction of a Mistake Previously Published." *Man*, (1960), 60:146.

1586 ———. "The Rules of Relationship Behaviour in One Variety of Primitive Warfare." *Man*, (1947), 47:108-10.

1587 ———. "Law and Force in Papuan Societies." AA, (1947), 49:244-59.

1588 ———. "Arapesh Warfare." AA, (1939), 41:22-41.

1589 FREEMAN, L.R. "Blood Debts of Savage New Guinea." *World Today*, (May 1924), 533-42.

1590 GARDNER, Robert. *Dead Birds*. (A film about tribal warfare among the Dani of western New Guinea.) Peabody Museum, Harvard University; Distributed: New York, Contemporary Films, 16mm.; 83 min.; sound; color; 1964.

1591 HADDON, A.C. "The Ingeri Head-hunters of New Guinea." *Internationales Archiv für Ethnographie*, (1891), 4:177-181.

1592 HEIDER, Karl G. *The Dugum Dani: A Papuan Culture in the Highlands of West New Guinea*. Chicago: Aldine, 1970, chapt. 3.

1593 HOGBIN, H. Ian. "Puberty to Marriage: A Study of the Sexual Life of the Native of Wogeo." *Oceania*, (1945/46), 16:185-209.

1594 ——. "The Father Chooses his Heir." *Oceania*, (1940/41), 11:1-39.

1595 ——. "Native Land Tenure in New Guinea." *Oceania*, (1939/40), 10:113-65.

1596 ——. "Social Reaction to Crime: Law and Morals in the Schouten Islands, New Guinea." JAI, (1938), 68:223-62.

1597 KOCH, Klaus-Friedrich. "Warfare and Anthropology in Jale Society." *Anthropologica*, (1970), 12:37-58.

1598 LARSON, Gordon F. *The Structure and Demography of the Cycle of Warfare among the Ilaga Dani of West Irian.* Department of Anthropology, University of Michigan, a Ph.D Major Research Paper, 1971.

1599 ——. "Warfare and Feuding in the Ilaga Valley." *Working Papers in Dani Ethnology*, v. 1. Holland: Bureau of Native Affairs, 1962.

1600 LORIA, Lamberto. "Notes on Ancient War Customs of Logea." In *British New Guinea Annual Report for 1894-5.* London, 1896, pp. 44-54.

1601 MATTHIESSEN, Peter. *Under the Mountain Wall.* (The Dani of New Guinea) New York: Viking, 1962.

1602 MEAD, M. "A Savage Paradigm." (A review of "Dead Birds," a film on tribal warfare by Robert Gardner.) *Film Comment*, (1964), 2:14-15.

1603 O'BRIEN, Denise. "Women's Warfare: Sorcery among the Kanda Valley Dani." Paper presented at the 68th annual meeting of the American Anthropological Association, New Orleans, 1969.

1604 POSPISIL, Leopold J. *Kapauku Papuan Economy.* New Haven: Yale University Press, 1963, pp. 29, 33, 40, 50, 58, 106, 130, 327.

1605 ——. "The Kapauku Papuans and Their Kinship Organization." Unpublished manuscript. New Haven: Human Relations Area Files, 1959, pp. 6, 8.

1606 ——. *Kapauku Papuans and Their Law.* New Haven: Yale University Press, 1958, pp. 15, 28-29, 90-93, 147-50, 159-63, 237-38.

1607 RAPPAPORT, Roy. "Ritual Regulation of Environmental Relations among a New Guinea People." *Ethnology*, (1967), 6:17-30.

1608 REAY, Marie. "Social Control amongst the Orokaiva." *Oceania*, (1953-54), 24:110-18.

1609 ROWLEY, C.D. "Culture Clash—Lime and Gunpowder." *South Pacific*, (1952), 6:513-15.

1610 SCHEFFLER, H.W. "Genesis and Repression of Conflict: Choiseul Island." AA, (1964), 66:789-804.

1611 SMARK, P. "The Dunas of the P-NG Highlands Fight for Fun." *Pacific Islands Monthly,* (1959), 2:61-63.

1612 THURNWALD, R. "Bei Kriegsausbruch im Zentralgebirge von Neu Guinea." *Koloniale Rundschau,* (1929), 38-43.

1613 TOWNSEND, G.W.L. "Psychological Warfare by New Guinea Natives." *Blackwood's,* (London, 1946), 259:395-98.

1614 Van der KROEF, Justus M. "Some Head-hunting Traditions of S. New Guinea" (1952), AA, 54:221-235.

1615 WATSON, J.B. "Tairora: The Politics of Despotism in a Small Society." *Anthropological Forum,* (1967), 2:53-104.

1616 WHITTING, John W.M. "The Frustration Complex in Kwoma Society." *Man,* (1944), 44:140-44.

1617 WILLIAMS, Francis B. *Orokaiva Society.* London: Oxford University Press, 1930, pp. 146-47, 163-71, 218, 312.

1618 WIRZ, Paul. "Kopfjagd und Trophaenkult im Gebiete des Papuagolfes." *Ethnology Anzeiger,* (1933), 3:201-3.

1619 ——. "Head-hunting Expeditions of the Tugeri into the Western Division of British New Guinea." TIJD, (1933), 73:105-122.

W.4 / New Zealand

1620 BEST, Elsdon. *The Maori.* Wellington: H.H. Tombs, 1924, v. 1:15, 64, 227, 240-42, 268, 275, 343, 354-55, 407, 440, 463.

1621 ——. *The Maori.* Wellington: H.H. Tombs, 1924, v. 2:90, 119, 144, 166, 227-99, 313-14, 332-45.

1622 ——. "Notes on the Art of War as Conducted by the Maori of New Zealand." JPS, (1902), 11:11-41, 47-75, 127-162, 219-46.

1623 ——. "Notes on the Art of War as Conducted by the Maori of New Zealand." JPS, (1903), 12:32-50, 65-84, 145-65, 193-217.

1624 ——. "Notes on the Art of War as Conducted by the Maori of New Zealand." JPS, (1904), 13:1-19, 73-82.

1625 BOHANNAN, Paul. "Review of Vayda, 1960, *Maori Warfare.*" JPS, (1961), 70:517.

1626 BUCK, Peter Henry (Te Rangi Hiroa). *The Coming of the Maori.* 2nd ed. Wellington: Maori Purposes Fund Board, 1952, pp. 276, 280-81, 346, 388-89, 396-400, 456, 460-61.

1627 CRANSTONE, B.A.L. "Review of Vayda, 1960, *Maori Warfare.*" *Man,* (1962), 62:144.

1628 DEWEY, A.G. "Review of Vayda, 1960, *Maori Warfare.*" AA, (1962), 64:407-8.

1629 FISCHER, H.I. "Review of Vayda, 1960, *Maori Warfare.*" *Sociologus,* (1962), 12:181-83.

1630 GRAHAM, G. "The Wars of Ngati-Huarere and Ngati-Maru-Tahu of Hauvaki Gulf." JPS, (1920), 29:37-41.

1631 GUDGEON, W.E. "Maori Wars." JPS, (1907), 16:13-42.

1632 ———. "The Toa Taua or Warrior." JPS, (1904), 13:238-64.

1633 KELLY, L.G. "Fragments of Ngapuhi History: The Conquest of the Ngareraumati." JPS, (1938), 47:163-72.

1634 ———. "Fragments of Ngapuhi History; Moremu-nui, 1807." JPS, (1938), 47:173-81.

1635 MANING, F.E. *Old New Zealand, A Tale of the Good Old Times; and a History of the War in the North Against the Chief Heke, in the Year 1845.* London: Bentley, 1876.

1636 NIHONIHO, T. "Uenuku or Kahukura: The Rainbow God of War." *Te Ao Hou,* (New Zealand, 1959), 7:50-53, 64-67.

1637 SCHWIMMER, E.G. "Warfare of the Maori." *Te Ao Hou,*(1961),36:51-55.

1638 SMITH, S. Percy. *Maori Wars of the Nineteenth Century.* Christchurch: Whitcombe and Tombs, 1910.

1639 VAYDA, A.P. "Maoris and Muskets in New Zealand: Disruption of a War System." *Political Science Quarterly,* (1970), 85:560-84.

1640 ———. "Maori Warfare." In Bohannan, *Law and Warfare,* (1967), 359-80.

1641 ———. *Maori Warfare.* Wellington: Polynesian Society Maori Monographs, No. 2, 1960.

W.5 /Australia

1642 BARCLAY, Arthur. "Life at Bathurst Island Mission." *Walkabout,* (1939), 8:12-19.

1643 BASEDOW, Herbert. *The Australian Aboriginal.* Adelaide: F.W. Preece, 1925, pp. 183-88, 273.

1644 CHASELING, Wilbur S. *Yulengor: Nomads of Arnhem Land.* London: Epworth Press, 1957, pp. 10-13, 22, 64, 77-79, 101-4, 110-11, 117-18, 125.

1645 HART, Charles W.M., and A. Pilling. *The Tiwi of North Australia.* New York: Holt, 1960, pp. 83-87, 174-75.

1646 ROTH, H. Ling. *The Aborigines of Tasmania.* London: Kegan Paul, Trench, Trubner, 1890, pp. 2, 18-19, 82, 87-94.

1647 SPENCER, Walter B. *The Arunta.* 2 vols. London: MacMillan, 1927, pp. 110-11, 443-53, 482, 506-7.

1648 STREHLOW, Theodor G.H. *Aranda Tradition.* Melbourne: Melbourne University Press, 1947, pp. 62-66.

1649 THOMSON, Donald F. *Economic Structure and the Ceremonial Exchange Cycle in Arnhem Land.* Melbourne: Macmillan, 1949, pp. 18, 39, 61-62, 76.

1650 TURNBULL, Clive. *Black War.* London: Cheshire-Lansdowne, 1948.

1651 WARNER, W. Lloyd. "Causes of War in Australia." PH, (1953), 507-8.

1652 ———. *A Black Civilization.* New York: Harper and Brothers, 1937, pp. 5, 27-28, 62-63, 86, 98, 110, 129, 155-90.

1653 ———. "Murngin Warfare." *Oceania,* (1930), 1:457-94.

1654 WEBB, T. Theodor. *The Aborigines of East Arnhem Land, Australia.* Melbourne: Methodist Laymen's Missionary Movement, 1934, pp. 17, 21.

1655 WHEELER, G.C. *The Tribe, and Intertribal Relations in Australia.* London, 1910.

Reference Matter

AUTHOR INDEX

Abel, T. 634
Abercombie, W.R. 915
Ackerknecht, E.H. 554, 555, 556
Adair, J. 1093
Adams, F.P. 945
Adams, M.N. 110
Adcock, F.E. 1394
Aitchison, T.G. 589
Ajayi, J.F.A. 1267, 1268
Albright, W.F. 806
Alland, A. 204, 205
Allardyce, W.L. 1524
Allen, W.E.D. 1286, 1395, 1396
Alm, J. 807
Almond, N. 111
Alter, J.C. 1003
Anderson, E.A.H. 1319
Anderson, F. 590
Anderson, R. 591
Andreski, S. 2
Andrezejewski, B.W. 1287
Andrzejewski, S. 635
Angell, R.S. 7, 248
Anthropova, V.V. 636
Apert, E. 206
Aragon, J.O. 230
Ardrey, R. 231, 249, 291
Armitage, C.H. 1270
Armstrong, J.M. 1163
Armstrong, P.A. 946
Arnold, R.R. 1004
Arnott, J. 1164
Aron, R. 3, 7
Aronson, D.R. 250
Asch, T. 367
Aufenanger, H. 592, 751

Austen, L. 471, 1578
Ayyangar, Sri M.A. 1427

Bacdayan, A.S. 366, 405
Bacon, R.F. 808
Baden, P. 406
Bailey, B.A. 773
Bajwa, F.S. 637
Baldus, H. 1165, 1166
Balfour, H. 809, 810, 811, 812, 813
Balikci, A. 1397
Ball, T.H. 955
Bandelier, A.F. 1132
Bannon, J. 1133
Barbey, G. 1288
Barclay, A. 1642
Barereba, S. 1579
Barker, G.T. 1525
Barker, J. 1167, 1168
Barros, A. 1169
Barry, D. 251
Bartell, G.D. 252
Barth, F. 1336
Bartlett, K. 1053
Barton, R.F. 472, 1497, 1498, 1499, 1500
Basauri, C. 1054
Bascom, W.R. 1428
Basedow, H. 1643
Basso, K.H. 1055
Basu, M.N. 4
Batchelor, J. 1429
Bates, O. 774
Bauer, I. 1337
Beach, W.W. 407, 1001
Beachey, R.W. 700
Beaglehole, E. 1056

Beals, A.R. 5, 252, 253, 286, 325, 326, 327, 356, 357
Bearsley, H.G. 638
Beattie, J.H.M. 1258
Beauclair, I. 701, 1526
Beaver, W.N. 511
Beemer, H. 639
Bell, F.L. 328, 1527
Benedict, B. 329
Benedict, M.J. 1367
Benedict, R. 6
Benson, M.G. 640
Benton, S. 775
Berkeley, G.F.H. 1259
Berkowitz, L. 254
Bernal Villa, S. 1170
Bernard, J. 7
Berndt, R.M. 255, 1580, 1581
Best, E. 641, 642, 1620, 1621, 1622, 1623, 1624
Bhaskaran, K. 1430
Bigelow, R. 207
Biocca, E. 1171
Birmingham, D. 232
Bittker, T.E. 148
Black, W. 519
Blackman, W.S. 1338
Blackwell, H.F. 1339
Blackwood, B. 1528
Blankenstein, M. van 408
Bledsoe, A.J. 1057
Blok, A. 1398
Bloom, L.B. 1058
Boardman, E.P. 112
Boas, F. 149, 916

111

TRIBAL NAME INDEX

DATE DUE

DEMCO 38-297